MONOGRAPHS OF THE
SOCIETY FOR RESEARCH IN
CHILD DEVELOPMENT

Serial No. 247, Vol. 61, No. 3, 1996

WHAT YOUNG CHIMPANZEES KNOW ABOUT SEEING

Daniel J. Povinelli
Timothy J. Eddy

WITH COMMENTARY BY
R. Peter Hobson
Michael Tomasello

AND A REPLY BY THE AUTHORS

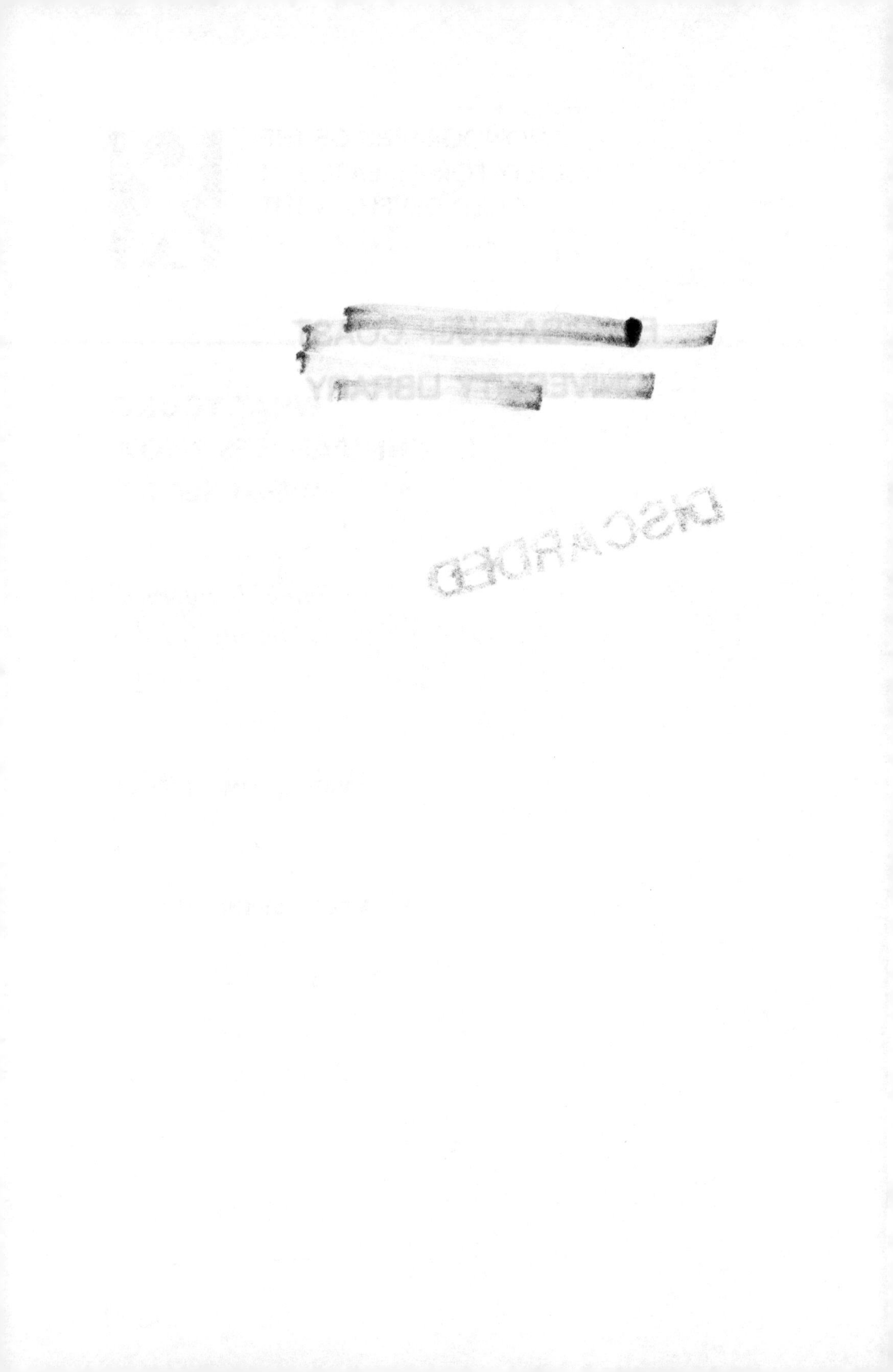

MONOGRAPHS OF THE SOCIETY FOR RESEARCH IN CHILD DEVELOPMENT
Serial No. 247, Vol. 61, No. 3, 1996

CONTENTS

ABSTRACT

POVINELLI, DANIEL J., and EDDY, TIMOTHY J. What Young Chimpanzees Know about Seeing. With Commentary by R. PETER HOBSON and MICHAEL TOMASELLO; and a Reply by DANIEL J. POVINELLI. *Monographs of the Society for Research in Child Development,* 1996, **61**(3, Serial No. 247).

Previous experimental research has suggested that chimpanzees may understand some of the epistemological aspects of visual perception, such as how the perceptual act of seeing can have internal mental consequences for an individual's state of knowledge. Other research suggests that chimpanzees and other nonhuman primates may understand visual perception at a simpler level; that is, they may at least understand seeing as a mental event that subjectively anchors organisms to the external world. However, these results are ambiguous and are open to several interpretations.

In this *Monograph,* we report the results of 15 studies that we conducted with chimpanzees and preschool children to explore their knowledge about visual perception. The central goal of these studies was to determine whether young chimpanzees appreciate that visual perception subjectively links organisms to the external world. In order to achieve this goal, our research incorporated three methodological objectives. First, we sought to overcome limitations of previous comparative theory of mind research by using a fairly large sample of well-trained chimpanzees (six to seven animals in all studies) who were all within 8 months of age of each other. In contrast, previous research has typically relied on the results of one to four animals ranging widely in age. Second, we designed our studies in order to allow for a very sensitive diagnosis of whether the animals possessed immediate dispositions to act in a fashion predicted by a theory of mind view of their psychology or whether their successful performances could be better explained by learning theory. Finally, using fairly well-established transitions in preschool children's understanding of visual perception, we sought to establish the validity of our nonverbal methods by testing predictions about how children of various ages ought to perform.

Collectively, our findings provide little evidence that young chimpanzees understand seeing as a mental event. Although our results establish that young chimpanzees both (*a*) develop algorithms for tracking the visual gaze of other organisms and (*b*) quickly learn rules about the configurations of faces and eyes, on the one hand, and subsequent events, on the other, they provide no clear evidence that these algorithms and rules are grounded in a matrix of intentionality. Particularly striking, our results demonstrate that, even though young chimpanzee subjects spontaneously attend to and follow the visual gaze of others, they simultaneously appear oblivious to the attentional significance of that gaze. Thus, young chimpanzees possess and learn rules about visual perception, but these rules do not necessarily incorporate the notion that seeing is "about" something.

The general pattern of our results is consistent with three different possibilities. First, the potential existence of a general developmental delay in psychological development in chimpanzees (or, more likely, an acceleration in humans) leaves open the possibility that older chimpanzees may display evidence of a mentalistic appreciation of seeing. Second, chimpanzees may possess a different (but nonetheless mentalistic) theory of attention in which organisms are subjectively connected to the world not through any particular sensory modality such as vision but rather through other (as-of-yet unspecified) behavioral indicators. Finally, a subjective understanding of visual perception (and perhaps behavior in general) may be a uniquely evolved feature of the human lineage.

I. RECONSTRUCTING THE EVOLUTION OF PSYCHOLOGICAL DEVELOPMENT

Had we conducted a search of our planet for some representative specimens of great apes and humans 20 million years ago, we would have come up empty-handed. The reason is simple: none of these species had evolved yet. However, at some point during the ensuing 15 million years, the ancestral populations from which humans and great apes emerged became reproductively isolated from each other. As a consequence, their descendants have been diverging in both their morphology and their behavior ever since. Figure 1 depicts the living descendants of this evolutionary radiation, along with other closely related taxa. Understanding the psychological consequences of the evolution of this group lies at the heart of our research agenda, a part of which we report in this *Monograph.*

We are particularly interested in the psychological evolution of this particular group of species for two reasons. First, we are convinced that we will never fully understand the true function of our own psychological structures until after we fully understand the timing and purpose of their origin. Thus, given that modern humans emerged from this group, exploring the similarities and differences among its living representatives is an obvious first step in understanding the timing of this psychological evolution. The second reason for focusing on this group is that the great ape–human clade may be the locus of rather profound evolutionary innovations in psychological development related to metacognition, resulting in cognitive abilities that (in the extreme case of recent human evolution) have made efforts such as this *Monograph* possible.

The great ape–human clade obviously represents only a few evolutionary strokes in a much larger, more diverse biological portrait. Although the species in this group may demand special attention, understanding their psychological evolution will require attention to more general relations that exist between descent and divergence that are characteristic of all evolving biological systems. In particular, it will require careful consideration of how psychological and morphological development and evolution may be inti-

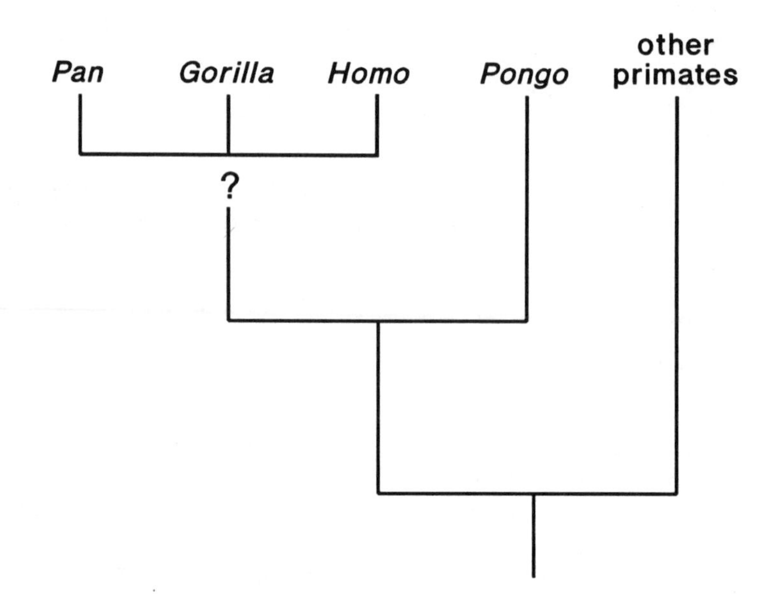

FIG. 1.—The evolutionary relationships of the great apes and humans, with other primates as an outgroup. The African apes (chimpanzees, gorillas) and humans are depicted as an unresolved trichotomy to reflect recent controversies among researchers attempting to reconstruct this phylogeny (see Marks, 1991, 1992).

mately related. Our research originates in this inherently interdisciplinary zone, drawing together elements from developmental and comparative psychology as well as evolutionary biology. Although comparisons of the psychology of human and nonhuman primates can be accomplished in the absence of such considerations, there is an underlying, theoretically driven motivation for such studies: reconstructing the timing and order of the evolutionary emergence of knowledge about the mind.

In an effort to provide a unifying framework that will contribute to the cross-fertilization among these different disciplines, this chapter sets the stage for a report of 15 studies that we conducted to determine what (if anything) young chimpanzees understand about the mental dimensions of visual perception. First, because our ultimate aim is to reconstruct the timing and order of the evolution of various aspects of metacognition, we briefly describe the logic of such reconstructions. Next, we assess the theoretical and methodological challenges to comparing the psychologies of different species. Finally, we describe recent comparative approaches to the development of theory of mind in monkeys, apes, and human children. Through these efforts, we show how the kinds of studies that we report are essential to the larger goal of understanding when an understanding of the mental world evolved.

RECONSTRUCTING PSYCHOLOGICAL EVOLUTION

The Logic of Evolutionary Reconstructions

Evolutionary biologists have understood the connection between descent and divergence at least since Darwin. Although he recognized that descent did not automatically imply divergence, Darwin argued that natural selection, acting on existing variation, tended to widen existing differences between populations across time. In principle, this idea of descent with modification poses no special problem: organisms with differing traits reproduce, offspring inherit those traits, and through a process of natural selection those organisms possessing traits positively linked to reproductive success contribute more offspring to subsequent generations. From this perspective, the strategy for tracking the evolutionary history of a particular trait (psychological or otherwise) would seem straightforward: find its initial point of appearance, and then trace out its subsequent line of descent.

This strategy can be applied to two general classes of evolutionary problems. First, the evolutionary history of specific traits can be specified. Second, identifying which species possess given traits provides the basis for what evolutionary biologists call phylogenetic reconstruction—the process of determining the exact structure of evolutionary trees (i.e., who is most closely related to whom). There are several approaches to reconstructing the evolution of various species (and their traits), but in general all of them rely critically on knowing in advance which species possess which traits. Once this has been achieved, inferences can be made about the most parsimonious account of the trait's evolutionary history.

Evolutionary reconstructions pose a number of challenges, but there are at least two types that are unique to attempts to reconstruct psychological evolution. One is methodological in nature, the other theoretical. We begin with the theoretical challenges because, were we unable adequately to counter them, overcoming the methodological difficulties would be pointless.

Do Phyletic Psychological Differences Exist?

The first theoretical challenge is simple: Is there any reason to suspect that species really possess different psychological characteristics? Evolutionary biologists accustomed to studying tightly canalized morphological and/or behavioral systems will find such a question astonishing, replying without hesitation that such differences must exist because diversity is the hallmark of all evolving biological systems. In contrast, comparative psychologists will not be similarly astonished, nor will they confidently offer such a glib defense of their discipline. Indeed, the inability of comparative psychologists to convince themselves that species differ with regard to fundamental psychological traits has been noted by many (e.g., Boakes, 1984; Burghardt & Gittleman, 1990; Hodos & Campbell, 1969; Lockard, 1971; Macphail, 1982, 1987; Wasserman, 1981). This difficulty in clearly identifying distinct psychological traits has its modern roots in Darwin's (1871/1982) insistence that continuity must exist in mental functioning between humans, their closest living relatives, and throughout the entire diversity of life. The challenge is a difficult one, for, even if we dismiss the extreme position that holds out for virtually no fundamental differences in psychological mechanisms across species (e.g., Macphail, 1982, 1987), we are still left with a lingering shadow of continuity that grips both behavioral and neurobiological approaches to understanding the evolution of primate cognition (Povinelli, 1993; Preuss, 1993).

There are two important replies to the objection that psychological evolution has, for some reason, escaped the general trend of branching diversity so characteristic of other biological systems. The first is that the domains in which psychologists have traditionally searched for qualitative phyletic differences are precisely those in which we should least expect to find them. In this sense, Macphail's (1987) quip that "causality is a constraint common to all ecological niches" exposes the soft underbelly of his more general claim that there are no differences in intelligence among vertebrates. Given that causality is a universal feature of biological environments, the types of general purpose learning mechanisms that Thorndike, Pavlov, Skinner, and others championed should be expected to be present in virtually all major taxonomic groups. But the universal presence of general learning mechanisms does not argue for the absence of specific psychological adaptations for specific behavioral patterns. Refocusing research efforts on ecologically relevant domains may lead to the detection of psychological differences in much the same way that comparative ethology successfully constructed phylogenies of distinct behavioral traits (Burghardt & Gittleman, 1990).

The final response to the argument that there are no qualitative psychological differences among species is that, although truly novel psychological innovations may be relatively rare, their rarity does not imply that they are unimportant. In the case of morphological evolution, there have been very few radical transformations in basic animal body plans, yet these core innovations constitute the basis for the classification of distinct phyla. So, too, radical alterations in psychological forms may occur relatively infrequently. But this is not to say that they do not occur at all. Indeed, the evolution of several such innovations may have already been detected by comparative psychologists (see Bitterman, 1975; Gallup, 1982; Rumbaugh & Pate, 1984). Thus, in addition to the detection of differing finely scaled psychological dispositions among species, large-scale transformations may also be detectable. In many cases, it is quite likely that the poor preservation of morphological and psychological traits (evolution of brain areas) in the fossil record may make such transitions appear "discontinuous" in living species (Povinelli, 1993).

Theory in Comparative Psychology

A second theoretical challenge to studying psychological evolution is that there is no theory in comparative psychology. Hodos and Campbell (1969) made precisely this claim in a landmark paper that blasted comparative psychologists for relying on an incorrect model of evolution as a basis for understanding the evolution of learning abilities. They charged that comparative psychology largely adhered to an outdated "phylogenetic scale" in which organisms inhabited various stages in a progressive evolutionary advance to higher forms of intelligence. This ascent up the phylogenetic scale reflected presumed progressive improvements in learning abilities. Hodos and Campbell (1969) strenuously objected to such efforts, noting that comparative psychology had relied on a model of evolution as a progressive process, as opposed to a process that results in radiating, branching diversity. An inspection of several major research programs of that era reveals the validity of their charges. But was the use of an inappropriate evolutionary theory an inherent flaw of comparative psychology, or did it merely reflect an aspect of the discipline gone awry?

In fact, not all comparative psychologists adopted the phylogenetic scale as their metaphor of psychological evolution. Hodos and Campbell (1969) clearly held out for the adoption of an explicitly neo-Darwinian framework to rescue the discipline. They noted that such a revamped comparative psychology could perhaps meet one of the central goals of any evolutionary science—reconstructing the timing and order of the appearance of evolutionary innovations. Since then, a number of authors have pointed out that a productive exchange of ideas is possible between those researchers who investigate the evolution of morphological development, on the one hand, and those who investigate the evolution of psychological development, on the other (Antinucci, 1989, 1990; McKinney & McNamara, 1991; Parker & Gibson, 1979; Povinelli, 1993). Indeed, McKinney and McNamara (1991) recently noted that the evolution of behavioral suites (and presumably their underlying neural substrates) may be tightly linked to the evolution of local morphological traits. Thus, in addition to the general point that some of the same principles of morphological evolution probably hold true for psychological evolution (e.g., brain evolution through heterochrony), it may also often be true that psychological and morphological evolution may be inextricably interwoven (for a hypothesized example in the case of the great apes and humans, see Povinelli, 1994b). In general, such an approach explicitly holds out for the detection of the evolution of unique brain areas in the course of primate evolution, which may be related to distinct morphological and behavioral adaptations (Preuss, 1993, 1995; Preuss & Goldman-Rakic, 1991a, 1991b).

An understanding of the patterns in which the psychologies of organisms diversify hand in hand with other aspects of their biology will ultimately provide a more complete understanding of whole organisms as units of selection. The evolutionary synthesis unified biology by showing how paleontology, systematics, and population genetics could all be forged on a common neo-Darwinian anvil. But organisms are more than protoplasm, muscles, and bones moving through space. They are also psychological systems that orchestrate their morphology into productive behavioral symphonies. A complete evolutionary synthesis must therefore incorporate the behavioral and brain sciences, including comparative psychology, for only comparative psychology can reveal how brain evolution opens up new behavioral strategies to be exploited. And, if one of the central efforts of psychology is to understand humans, then nowhere will the reward be greater than in the context of the evolution of the great ape–human clade. The specific research agenda of the studies reported in this *Monograph* represents one such effort.

Methodological Challenges of Comparing Psychologies

Perhaps the most common objection to traditional methods of comparative psychology is that its measurements are derived from laboratory tasks that are ecologically irrelevant and hence that the resulting comparisons are evolutionarily irrelevant. However, it is important to note that the ecological neutrality of traditional learning tasks was neither an accident nor an oversight. Indeed, if it is true that historically much research in the field has been confined to experimental situations that species would never encounter in nature, this is precisely because researchers planned it so. Comparative psychologists have traditionally been interested in neutralizing surface species differences, believing that such a method was the only way to derive equitable measures across various taxa (Seligman, 1970). Although this has changed dramatically in the last two decades, the questions that comparative psychology traditionally addressed concerned underlying processes, not ethological questions about the relation between species-typical repertoires or morphological adaptations and behavior. Yet this is not to say that researchers believed that the capacities measured were ecologically irrelevant; quite to the contrary, the abilities measured were presumed to be the principle mechanisms by which organisms modified their behavior in relation to a changing environment. Viewed in this context, the design of species-neutral tasks was a means of teasing apart extraneous species differences from "real" psychological ones. Thus, the design of ecologically liberated tasks was considered to be a crowning victory for comparative psychology, not its Achilles heel.

Clearly, there is merit to both the criticism of comparative psychology's historical emphasis on species-neutral tasks and the discipline's subsequent defense of them. An examination of the results of more recent research programs utilizing ecologically derived predictions about species differences in cognitive abilities ("constraints" on learning) reveals the potential benefits of such an approach (Cole, Hainsworth, Kamil, Mercier, & Wolf, 1982; Gaulin, 1992; Kamil, 1978; Shettleworth, 1975). Such researchers have provided strong, logically coherent defenses of turning to evolutionary biology as a means of generating predictions about species differences in cognition. Indeed, this approach has been cast as a means of rejuvenating comparative psychology (Kamil, 1983, 1984; Kamil & Yoerg, 1982). However, equally striking are phyletic differences obtained from comparative research originating from very little in the way of ecologically based predictions (Bitterman, 1975; Gallup, 1970; Rumbaugh & Pate, 1984). Given that both approaches have succeeded in yielding intriguing evidence of quantitative and qualitative species differences, what is needed is a framework that can embrace both of these strategies.

A step back from both sets of examples may help reveal the general outline of such a framework. On the one hand, both types of research strategies employ laboratory-based tests that even in the best circumstances only weakly resemble the ecologies in which the species actually evolved. For example, Gaulin's (1992) ingenious research on sexual dimorphisms in spatial intelligence in several species of microtine rodents involved, in part, running subjects through classic rat mazes. Kamil's research has often involved testing birds in Skinner boxes (both large and small). Clearly, then, the ecological *validity* of a task obviously has little to do with the ecological *appearance* of the task. The true element that separates both types of research programs from the traditional research programs of comparative psychology seems to be the search for domain-specific intelligences (see Kamil & Yoerg, 1982). To be certain, a priori predictions derived from evolutionary theory offer an important and increasingly indispensable framework for knowing where to look for phyletic psychological differences. However, research efforts following up on serendipitous discoveries may yield data that can later be assimilated into an evolutionary framework, which in turn can lead to equally testable predictions for further research.

Many researchers have balked at investigations of species differences, noting that such an approach invariably leaves researchers in the awkward position of attempting to prove the null hypothesis by showing that certain species do not possess a certain capacity or trait that is possessed by others. In a classic effort to stake out comparative psychology's theoretical and methodological territory, Bitterman (1960) detailed many of the inherent difficulties in trying to determine the existence of species differences. Part of his solution was systematically to vary experimental conditions across species. This was done in order to demonstrate that, no matter how the experimental situation was altered to control for hypothesized species differences unrelated to the phenomenon in question (e.g., motivation), some species would still perform differently than others. Among others, Kamil and Yoerg (1982) have chastised such an approach as being "boring" and, worse still, leaving "the experimenter in the untenable position of attempting to prove the null hypothesis" (p. 349). However, what researchers often overlook is that hypotheses about the absence of (or difference in) a particular psychological process are every bit as falsifiable as hypotheses that predict the presence of a given ability (Povinelli, 1993). Thus, researchers are not really in the position of attempting to prove the null hypothesis; rather, they are simply in the position of trying to falsify it.

A classic case in point is the long-standing problem of distinguishing between qualitative and quantitative differences in psychological abilities. Some researchers prefer to view this problem as a semantic one—a "non-problem" created by insufficiently precise definitions of either the particular traits under study or the descriptors *qualitative* and *quantitative.* However, a careful dissection of the problem reveals that it is a deep, conceptual one, involving questions of function versus process as well as the role of development in the evolutionary process (for a more detailed discussion, see Povinelli, 1993). Nonetheless, progress comparing the psychologies of nonhuman primates and humans need not await a definitive conceptual resolution of this problem; local, domain-specific investigations concerning phylogenetic (or developmental) differences can proceed through the standard scientific process of hypothesis testing. For example, Gallup (1970) demonstrated phyletic differences in the capacity for organisms to recognize themselves in mirrors and, on the basis of his data, hypothesized that the capacity was restricted to the great apes and humans.

Gallup's hypothesis (along with its assumed importance for the self-concept) has led researchers to develop a wide variety of ingenious (albeit unsuccessful) methods to demonstrate self-recognition in species outside the great ape–human clade. Indeed, the difficulty in demonstrating the phenomenon in other species led primatologists to attempt to teach monkeys, and behaviorists to teach pigeons, to recognize themselves in mirrors (Anderson, 1984; Epstein, Lanza, & Skinner, 1981; Itakura, 1987a, 1987b). Yet none of these efforts need to be seen as attempts to prove the null hypothesis. Rather, they can be viewed as attempts to falsify the hypothesis of phyletic differences in the capacity. More generally, this same reasoning applies to the entire field of cognitive development. The argument that species differences cannot be investigated because they inherently lead researchers to attempt to prove the null hypothesis would logically apply with equal force to the field of developmental psychology and thereby invalidate any claim of age-related differences in cognitive capacities (Goldman-Rakic & Preuss, 1987; Povinelli, 1993).

Conversely, the "successful" demonstration of the presence of an ability (i.e., of an organism's ability to detect *x* and do *y*) may turn out later to be unrelated to the causal process originally hypothesized. Third variables that systematically varied with the independent variables manipulated by the researcher may have caused the results. Thus, it is simply incorrect to claim that investigating species differences somehow uniquely places investigators in the role of trying to prove the null hypothesis.

COMPARING PSYCHOLOGICAL DEVELOPMENT

One research strategy in particular holds great promise for minimizing the methodological difficulties described above: comparing psychological development across species. The roots of such an approach can be traced back at least to Darwin (see the review by Parker, 1990). However, one of the earliest advocates of a broadly based comparative developmental approach was Alison Jolly (1964a, 1964b), who argued for the potential utility of applying Piagetian stage theory to studies of nonhuman primates. She noted that certain aspects of the ways in which adult lemurs responded to objects seemed to mirror certain kinds of reactions of infant humans during their sensorimotor development. Parker and Gibson (1979) elaborated on this approach and provided an initial attempt to unite it with mainstream evolutionary biology. There has since been a recent blossoming of research in this area, most of which has drawn on the Piagetian approach first suggested by Jolly over 30 years ago (see the contributions to Parker & Gibson, 1990). Although such an approach continues to hold great promise in comparing psychological development, our interests are not in establishing the utility or validity of a Piagetian framework. Instead, we ask how a comparison of developing psychologies can assist researchers in overcoming the methodological difficulties historically associated with comparative psychology in general.

Compelling demonstrations of developmental commonalities among closely related species can provide much needed confirmation of the underlying conceptual validity of the comparisons, for, if researchers are able to sketch the common features of development within a particular clade, then ontogenetic deviations across species (i.e., species differences) can be interpreted in an explicitly evolutionary light. Species differences in adult organisms no longer appear to exist miraculously, perhaps as mere artifacts of the demands of ecologically irrelevant tasks; rather, they can be seen as emerging from evolutionary innovations in ontogeny. For example, if we were able exhaustively to catalog the timing and order of the major cognitive developments for all the living species of the great ape–human clade, this would provide the necessary data from which we could determine which features are primitive for the clade and where in the various lineages' ontogenetic programs innovations have occurred.

This larger goal has been under way for quite some time. Just to select a particularly elegant series of studies, a recent comparative research program on sensorimotor development in macaques, capuchins, gorillas, chimpanzees, and humans offers an excellent example of how finely scaled investigations of cognitive development across species can implicate certain evolutionary processes in operation (Antinucci, 1989, 1990; Potì & Spinozzi, 1994). Although this particular set of studies has the drawback of not comparing all species within a closely related group and has (by necessity) relied on very small sample sizes, it nonetheless represents an impressive effort to understand how various common aspects of sensorimotor development are affected by alterations in the timing of other ontogenetic programs related to physical maturation (Antinucci, 1990). Povinelli (1994b) has speculated on how their results can provide insights into other puzzling aspects of cognitive development within the great ape–human clade.

Comparisons of psychological development within the great ape–human clade offer an ideal method of determining which features of human cognition are exclusively derived in the human lineage. For example, by carefully comparing the psychological development of chimpanzees, orangutans, gorillas, and humans, we will ultimately be able to specify where in their development they diverge, in which directions, and ultimately for what reasons. As we have seen, the preferred method to date for establishing common features of psychological development among members of this clade has been through the application of Piagetian stages to the various species (see Parker & Gibson, 1990). Although such work is critically important, our own research interests lie elsewhere. In the following sections, we provide a very general overview of the psychological domains in which we are interested in developing evolutionary reconstructions: knowledge about the mental world.

EVOLUTION OF THEORY OF MIND

In brief, our broad research agenda is to determine whether there have been psychological innovations during the evolution of primate ontogeny in the domain of what Premack and Woodruff (1978b) referred to as *theory of mind.* By *theory of mind,* Premack and Woodruff were referring to the beliefs that humans (at least) possess about their own and others' mental states. Turning behaviorism on its head, they took their theory of mind concept and focused their attention on chimpanzees, asking not simply whether chimpanzees possessed mental states but whether they were capable of making inferences about the mental states of others. "A system of inferences of this kind," they observed, "may properly be viewed as a theory because such states are not directly observable, and the system can be used to make predictions about the behavior of others" (Premack & Woodruff, 1978b, p. 515).

In a series of tests with a 16-year-old female chimpanzee named Sarah, they sought to determine whether she would spontaneously attribute intentions to others. They constructed a series of simple videotapes depicting human actors struggling to solve staged problems (e.g., a human trying to grasp a banana suspended just out of reach), and Sarah was given ample opportunity to observe them. The videotapes were then played back for her, and she was presented with pairs of still photographs, one of which represented the solution to the implied problems faced by the actors on the tapes. Premack and Woodruff reported that on a variety of such problems Sarah correctly selected the solution photograph on her first trial. They interpreted this as evidence that she "recognized the videotape as problem, understood the actor's purpose, and chose alternatives compatible with that purpose" (p. 515). After conducting these tests, as well as several control tests, to rule out less demanding interpretations of Sarah's behavior, Premack and Woodruff concluded that chimpanzees possess a theory of mind, at least insofar as attributing intention is concerned.

Several authors who commented on Premack and Woodruff's theory of mind paper noted that a more compelling demonstration of the presence of a theory of mind could be obtained by demonstrating that an organism understood the representational property of belief, in particular, by demonstrating that an organism could represent the false beliefs of others (Bennett, 1978; Dennett, 1978; Harman, 1978; see also Lewis, 1969). Premack and Woodruff (1978a) poked fun at the idea, but developmental psychologists were quick to seize the initiative and began to explore the development of young children's understanding of false belief (Wimmer & Perner, 1983). Other developmental psychologists began using the term *theory of mind* as well, for example, in investigations of the development of young children's use of language referring to mental states (Bretherton & Beeghly, 1982).

However, long before these efforts, there had been a less widespread interest in similar phenomena by several developmental psychologists. Piaget's (1932) early interest in the attribution of moral responsibility by young children led a number of researchers to look more carefully at what young children understand about intentions (Berndt & Berndt, 1975; Chandler, Greenspan, & Barenboim, 1973; Gutkin, 1972; Imamoğlu, 1975; Irwin & Moore, 1971; Karniol, 1978; King, 1971; Shultz & Shamash, 1981; Shultz, Wells, & Sarda, 1980; Smith, 1978). Similarly, Piaget and Inhelder's (1956) claims about egocentrism in the domain of visual perspective taking led a number of researchers to investigate more thoroughly the ability of young children to see the world through the eyes of others (Borke, 1971; Fishbein, Lewis, & Keiffer, 1972; Flavell, 1974; Flavell, Everett, Croft, & Flavell, 1981; Lempers, Flavell, & Flavell, 1977; Liben, 1978; Masangkay et al., 1974).

From these two origins, theory of mind has become one of the most researched areas in developmental psychology. Owing to the sheer number and scope of research efforts into children's theory of mind, in this chapter we can do little better than allude to some representative examples of this work. Investigations of young children's developing understanding and use of mental state terminology, false belief, deception, sources of knowledge, visual perspective taking, intentions and desires, the distinction between the appearance and the reality of objects, events, and emotions, pretend play, stream of consciousness, and thinking, all these can now be embraced under a loose usage of the term *theory of mind* (Astington & Gopnik, 1991; Bartsch & Wellman, 1995; Chandler, Fritz, & Hala, 1989; Flavell, 1988; Flavell, Green, & Flavell, 1993, 1995; Frye & Moore, 1991; Harris, 1989, 1991; Lillard, 1993; Perner, 1991; Wellman, 1990; see also the recent overview of major texts in the field by Moses & Chandler, 1992). Although not all authors agree on what constitutes evidence of an understanding or a comprehension of particular mental constructs (such as intention and belief), the field can be thought of as being united around a central question: At what age do children display evidence of understanding the existence of various aspects of the unobservable life of the mind?

Prior to Premack and Woodruff's (1978b) report, several researchers were exploring what animals might (or might not) know about the mind. Gordon Gallup (1970, 1975) explored the capacity of nonhuman primates to recognize themselves in mirrors. After discovering that the capacity appeared to be limited to the great ape–human clade, he argued for its relation to the possession of an explicit self-concept. Later, in what can be considered as the most ambitious attempt to date to provide a coherent theoretical framework for modern comparative psychology, Gallup (1982, 1983, 1985) offered a broader hypothesis about the potential restriction of theory of mind–like capacities to those species that displayed evidence of self-recognition in mirrors. From a very different perspective, Emil Menzel (1974) observed young chimpanzees engaging in what could be construed to be instances of the animals intentionally deceiving one another in the context of competitive situations that he created for them. Likewise, Frans de Waal (1982) described a long-term study of a group of chimpanzees and offered a captivating interpretation of chimpanzees as a Machiavellian species, grounded in political dominance struggles. However, these are notable exceptions. By and large, modern comparative psychology had little interest in what animals might know about mental states per se before Premack and Woodruff's report, and little interest was generated immediately after it (for other exceptions, see Cheney, Seyfarth, & Smuts, 1986; Kummer, 1982; Smuts, 1985).

In recent years, however, comparative psychology has returned in full force to the question of what animals know about the mind. The first entry point for this return built on de Waal's (1982) view of chimpanzees as political animals. To a large extent, these discussions have used anecdotes of deceptive behavior as the focus of discussion (Byrne & Whiten, 1985, 1991; Whiten, 1993; Whiten & Byrne, 1988). However, as sophisticated as such maneuverings appear, the ability of anecdotes to determine one way or the other what the animals' appreciate about mental states is limited (Bernstein, 1988; Burghardt, 1988; Heyes, 1988, 1993; Premack, 1988). Perhaps the most salient critiques of this approach are those that note that learning is not a phenomenon restricted to the laboratory. Premack (1988) was blunt: "How many 'trials' go into producing the anecdotes that are reported from the field? Since this is rarely known, readers are led to indulge their ignorance and draw romantic conclusions" (p. 171). Premack's point is not that the anecdotes themselves are unreliable but rather that their validity is always open to doubt; thus, replacing a single anecdote with multiple anecdotes does not enhance the validity of the inference to mental state attribution (see also Burghardt, 1988). Thus, as interesting as are the dozens of anecdotes collected and collated by Byrne and Whiten, they bring us no closer to answering the fundamental question implicitly posed by Edward Thorndike's discovery of the law of effect: Do animals understand social interactions solely in terms of behavior or in terms of both behavior and mind?

The second entry point for comparative psychology's return to the question of mental state attribution was laboratory investigations of the type advocated by Premack and Woodruff (Cheney & Seyfarth, 1990b; Povinelli, Nelson, & Boysen, 1990, 1992; Povinelli, Parks, & Novak, 1991, 1992; Premack, 1988; Woodruff & Premack, 1979). In contrast to the anecdotal approach of Whiten and Byrne, these investigators have explored the question of whether nonhuman primates reason about the mind by using experimental techniques explicitly designed for that purpose. To date, these investigations have yielded little evidence that monkeys (macaques, at least) reason about mental states, whereas there is at least suggestive evidence that chimpanzees might. As we will see in the next chapter, however, much of this research suffers from a similar problem as the anecdotal approach in that it has not relied on sensitive enough diagnoses of behavior, thus providing ample opportunity to confound learning with mental state attribution.

Povinelli (1993) has recently outlined a framework for an experimental comparative research program explicitly designed to determine when, and in what order, various aspects of mental state attribution evolved. This framework is not new but rather draws on previous work in the field of comparative psychological ontogeny, developmental psychology, and evolutionary biology. In particular, using the general nonverbal experimental techniques outlined by Premack and Woodruff (1978b), and applying them to the questions that have emerged from investigations of theory of mind in young children, it is now possible to determine whether theory of mind represents a psychological innovation unique to the human lineage or whether it is a more primitive innovation, perhaps one that evolved sometime after the divergence of the great ape–human lineage from the other primates. This latter view would be compatible with Gallup's (1982, 1983, 1985) model of the evolution of mental state attribution. Additionally, it is possible that the psychological innovations responsible for theory of mind dispositions were not, in fact, a single innovation at all but rather evolved in a number of fairly discrete steps (Povinelli, 1991, 1993; Premack, 1988). Thus, without recourse to "recapitulation," it is quite possible that transitions in theory of mind dispositions identified by developmental psychologists represent the retention of discrete ontogenetic innovations during the course of primate evolution. Figure 2 represents two extreme ends of this spectrum of possibilities.

RECONSTRUCTING THE EVOLUTION OF THEORY OF MIND

It is clear from the above considerations that reconstructing the evolution of theory of mind will proceed through three distinct phases. To begin, researchers must use the methods of comparative psychology to identify which species possess which aspects of mental state attribution and at what point in development. Although such an effort has begun, the results to date are considerably less than definitive (Cheney & Seyfarth, 1990b; Povinelli, 1993; Whiten, 1993). The second step will be for researchers to use the methods of phylogenetic reconstruction to infer what the likely features of theory of mind were in each common ancestor of the great ape–human clade. Once this reconstruction has occurred, the exact timing and order of

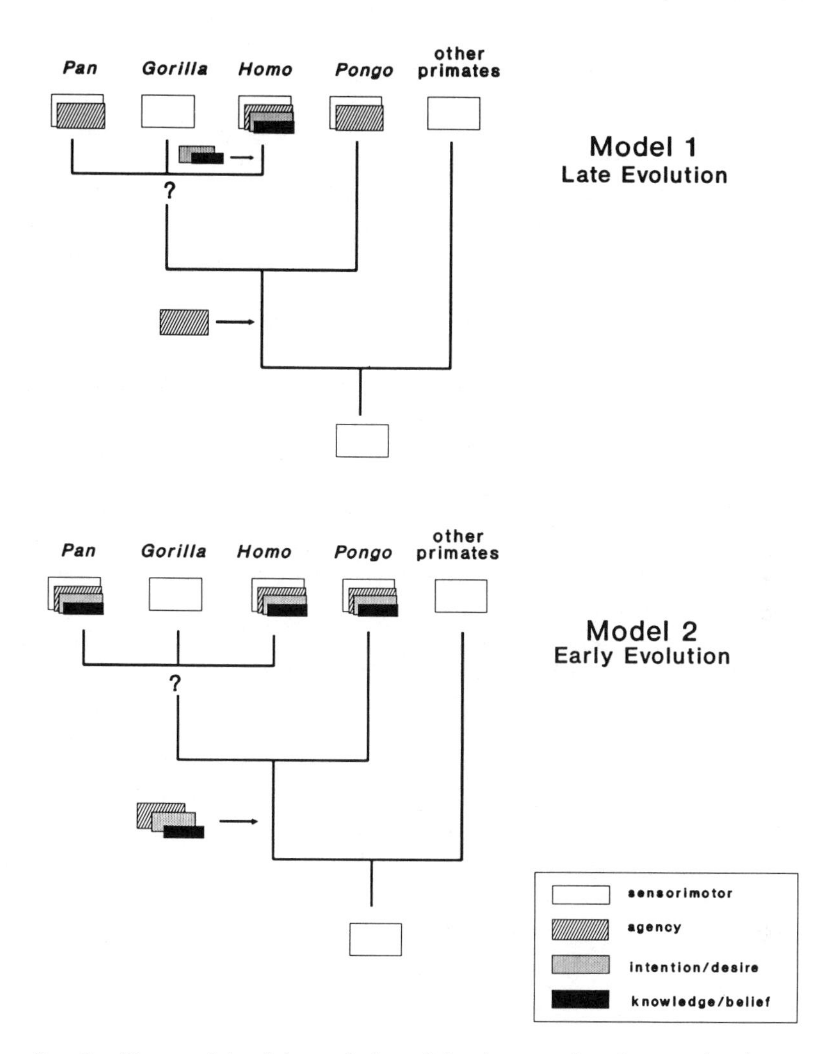

FIG. 2.—Two models of the evolution of developmental pathways related to mental state attribution. In Model 1, the capacity for conceiving of agency evolved in the ancestor of the great ape–human clade and is thus shared in most of the living descendants of this group. Unique evolution in the human lineage, including most conceptual capacities underlying the understanding of mental states per se, occurred later. In Model 2, the bulk of the developmental pathways governing mental state attribution evolved in the ancestor of the great ape–human clade and are thus present in most of the descendants. In both models, gorillas are represented as having undergone unique evolution resulting in the reversal of certain character states (see Antinucci, 1989; Potì & Spinozzi, 1994; Povinelli, 1993; for general data on unique changes in physical growth and maturation rates in gorillas, see McKinney & McNamara, 1991; Watts & Pusey, 1993). A detailed hypothesis about the secondary loss of certain cognitive pathways in gorillas is provided by Povinelli (1994b). There may be unique innovations in other lineages as well. These two models simply represent fairly extreme ends on a spectrum of possibilities of the exact timing of the evolution of mental state attribution.

the emergence of each of the features will be known, and it will be possible to choose between the models outlined in Figure 2 or to implicate an alternative model. The final, and most difficult, step will be for researchers to determine why these capacities evolved in the order and with the timing that they did. This final phase of the research agenda will explicitly involve developing evolutionary scenarios that explain why these abilities emerged when they did. These three phases of reconstructing theory of mind evolution need not necessarily proceed in the order outlined here. For example, adaptive scenarios can be posited even in the face of our currently poor understanding of which species develop a theory of mind. However, those who argue strenuously for particular scenarios before a reasonable data set emerges run all the risks inherent in putting the cart before the horse.

The studies reported in this *Monograph* fall squarely into the first phase of this overall research agenda: identifying which species possess which traits. Indeed, we ask a very narrow question: Do chimpanzees (and young chimpanzees at that) appreciate how visual perception (seeing) subjectively links organisms to the external world? In other words, do chimpanzees interpret seeing as a mental event or as an exclusively behavioral act? We can be certain that humans appreciate the mental dimension of visual perception, and the work of Flavell and his colleagues has established that even very young humans understand at least this simple fact about seeing (e.g., Lempers et al., 1977). In the next chapter, we review the theoretical and empirical work in this area for both humans and nonhuman primates. We use this information to construct two frameworks for understanding what young chimpanzees might know about seeing: one arguing that they understand it strictly as a behavioral event, the other arguing that they understand it as both a behavioral and a mental event. In Chapters III, IV, and V, we report on 15 experiments that we conducted with young chimpanzees and children, all of which were designed to assist us in choosing between these alternative accounts of young chimpanzees' knowledge about seeing.

II. UNDERSTANDING VISUAL PERCEPTION

An organism's perceptual systems can be thought of as its psychological anchors to the external world. The primary mechanisms of perception (hearing, touch, olfaction, taste, vision) provide organisms with a means to monitor various dynamic aspects of the physical world: for example, sound, temperature, texture, chemical content, and visual appearance. Organisms that use their perceptual systems to track these features of the world better than their competitors are likely to outreproduce them. Thus, there can be little doubt that these mechanisms of perception evolved in the course of metazoan evolution precisely because they provided information about objects and events intimately linked with reproductive success. From a related perspective, it should not be surprising to discover that individual species have evolved perceptual systems that are finely tuned to detect the resources and events that are of greatest reproductive consequence for them. One such evolutionary refinement has clearly occurred in highly social species in the coevolution of *visual* perception and the behavior of other organisms and, in particular, the use of one's own visual sensory system to monitor the visual sensory system of others (e.g., Chance, 1967; Fehr & Exline, 1987). However, there are several distinct aspects of the visual systems of others that organisms may or may not monitor, and these can be described as differing degrees of knowledge about the mental dimensions of visual perception.

Knowledge about visual perception can be processed on at least three levels of comprehension. First, it is possible for organisms to attend and respond to other organisms' eyes without appreciating anything at all about their function or epistemic significance. This level of information processing about vision is illustrated by the widespread sensitivity to the presence of eyes in birds, reptiles, fish, and mammals (e.g., Burger, Gochfeld, & Murray, 1991; Burghardt & Greene, 1988; Gallup, Nash, & Ellison, 1971; Ristau, 1991; for a review, see Argyle & Cook, 1976). In addition, a sensitivity to eyes (and eye-like patterns) appears very early in human development, leading to the suggestion that such sensitivity might be tightly canalized (Johnson & Morton, 1991; Spitz, 1965; Spitz & Wolf, 1946; for a recent "modular" theory of this canalization, see Baron-Cohen, 1994). Although such examples demonstrate that many species and very young infants are sensitive to the eyes, they in no way provide unique evidence that they understand any of the mentalistic properties or consequences of visual perception.

A second aspect of visual perception that organisms can potentially understand is that vision subjectively links individuals to the world. That is, not only does vision, in truth, have the effect of focusing the subjective attention of individuals on specific aspects of the world, but organisms can also potentially know this fact. The common metaphor of "eye contact" highlights this property well—although the eyes do not literally make contact with the external world, we nonetheless interpret eye gaze as linking the perceiver to that which is perceived. At this level of information processing, visual perception may be equated with a subjective connection to the external world (Flavell, 1988). Although both the developmental and the evolutionary emergence of this understanding of visual perception is unclear, recent work with human infants is beginning to provide evidence of an early understanding of the attentional focus of others, although it is not at all clear whether these infants understand the role of visual perception in this attentional process (see Baldwin & Moses, 1994).

Finally, at the most complex level that we will consider here, organisms may understand the mental consequences of seeing, that is, the role that visual perception plays in creating internal, unobservable states of knowledge (or belief). In short, they may understand that visual perception in themselves and others (seeing) leads to knowledge formation (knowing) (Gopnik & Graf, 1988; Wimmer, Hogrefe, & Perner, 1988). This can be viewed as a more sophisticated understanding of the mental dimension of visual perception than understanding merely that vision connects organisms to the world. If an organism interprets visual perception as having consequences for internal mental states, it should be able to attribute a diversity of states of knowledge to others on the basis of the extent of visual contact with a situation. In this sense, even long after visual contact is broken, someone who has seen an event is in a very different epistemological position than someone who did not.

Having outlined several levels at which visual perception can be processed or understood by organisms, we now turn our attention exclusively to the two that have an unquestionable connection to theory of mind development. Thus, below we review the developmental and comparative evidence for the emergence of the understanding that visual perception subjectively connects organisms to the world and an understanding of it as a knowledge acquisition device. This review sets the theoretical stage for asking several specific questions about what other species, and in particular chimpanzees, know about seeing.

UNDERSTANDING VISUAL PERCEPTION AS ATTENTION

Although there is little debate that infants selectively attend to eyes at a very young age, there is considerably greater disagreement about whether this qualifies as evidence of intersubjectivity or an awareness of the presence of the minds that presumably lie behind those eyes. A number of researchers have argued that the development of mutual gaze, or gaze alternation, between infant and mother sets the stage for the infant's entry into the mental world. For example, Trevarthen and Hubley (1978) argue for intersubjectivity at 3–4 months and secondary intersubjectivity at 7–9 months, in large part on the basis of the development of mother-infant eye gaze patterns. Indeed, eye contact has been held as a defining criterion for the presence of intentional communication (Bretherton & Bates, 1979). Still others have demonstrated that by 9 months of age human infants will scan a person's eyes immediately following an ambiguous action, seeming to demonstrate that they understand that "people are goal-directed in their action" (Phillips, Baron-Cohen, & Rutter, 1992, p. 382). Other researchers doubt such early sensitivity to eyes and eye gaze as evidence of a subjective understanding of others, arguing that such behaviors may merely be learned rules with specific payoffs (e.g., following another's line of gaze results in useful information about the environment; see Butterworth & Jarrett, 1991; Scaife & Bruner, 1975).

Studies of social referencing in infants (the ability to establish a connection between another's attentional focus and behavioral/emotional reactions) might also be viewed as evidence that infants understand something about the subjective (mental) side of visual perception (Campos, 1983; Feinman, 1982; Feinman, Roberts, Hsieh, Sawyer, & Swanson, 1992). For example, Baldwin (1991) recently provided evidence that by 16–18 months infants may be able to use another's attentional focus (which includes visual gaze) in order to decipher the referent of nonsense words. Baldwin and Moses (1994) conclude that such research demonstrates that sometime between 12 and 18 months infants come to understand the attentional focus of another person as "something like a psychological spotlight that can be intentionally directed at external objects and events" (p. 151). In general, regardless of the exact developmental timing of this ability, this tradition of research has attempted to isolate a particular form of knowledge about attentional focus: that it is *intentional* in the sense that it refers to (or is "about") something and thus signifies the mental state of attention. However, it is important to emphasize that there is no evidence that infants of this age understand the exact role that the eyes play in grounding attentional focus.

Other researchers have investigated young children's knowledge about visual perception specifically, asking at what age children understand what others can and cannot see. In a series of reports, John Flavell and his colleagues demonstrated that children of 2½ years and older can accurately judge whether another person can see an object and can even produce situations that result in depriving others of visual contact with an object on request (Flavell et al., 1981; Flavell, Flavell, Green, & Wilcox, 1980; Flavell, Shipstead, & Croft, 1978; Lempers et al., 1977). Lempers et al. (1977) conclude that by 2½ years of age children appear to understand the role of another's eyes in seeing. Flavell (1988) interprets this as evidence that young children understand that visual perception "cognitively connects" people (and other organisms) to the external world. Of course, it is possible that the studies of social referencing and attentional focus reviewed above represent an even earlier instance of this kind of understanding of how perception through the eyes subjectively connects organisms to the external world, but again there is no evidence that these young infants specifically understand the role of the eyes in establishing attentional focus.

From a comparative perspective, evidence of careful monitoring of eye gaze during social interactions by some nonhuman primates could be interpreted as evidence that they interpret visual perception as indicating something like the mental state of attention (Gómez, 1991a, 1991b; Menzel, 1971, 1974; Menzel & Johnson, 1976; Whiten, 1991, 1993). For example, in studying the ontogeny of an infant gorilla, Gómez (1991a, 1991b) reports that at first the gorilla attempted to move people toward desired goals by pushing or dragging them in order to use them as a ladder (for similar descriptions of young chimpanzee behavior, see Köhler, 1927). However, the gorilla later attempted to lead people by the hand to the goal while alternately gazing between their eyes and the desired goals. Gómez (1991b) interprets this as evidence of the gorilla's mentalistic understanding that "attentional contact between subjects" is needed in communicative interactions. He argues that such careful attention to eye contact is evidence that the organism knows that an individual must visually attend to its gestures in order for them to be effective. However, the extent to which it is evidence of a *mentalistic* understanding of visual perception remains uncertain. Gómez (1991b) acknowledges that the evidence to date demonstrates only that gorillas perceive a link between the overt behavioral manifestations of visual gaze and the actions of others, not that they represent this connection in any mental fashion. Thus, evidence of either mutual gaze or gaze monitoring, by itself, demonstrates only that organisms have learned that eyes provide important predictive clues about the behavior of others.

This problem can apply with equal force to some studies of social referencing and joint attention in human infants. It is quite possible that organisms may monitor the eyes of others without understanding that they are cognitively connected (via visual perception) to the situation at hand. In other words, eyes may develop (or even initially possess) a high valence for organisms independent of an understanding of their linkage to the internal mental state of attention.

UNDERSTANDING VISION AS A KNOWLEDGE ACQUISITION DEVICE

Beyond simply understanding that vision connects the subjective experience of organisms to the world, humans, at least, understand that visual perception plays a causal role in creating states of knowledge. The fact that an organism appreciates the attentional property of "seeing" (i.e., knowing that visual perception connects organisms to the world) in no way guarantees that it appreciates the mental consequences of that perception. For example, although the research of Lempers et al. (1977) demonstrated that even 2-year-olds could determine what someone could or could not see, parallel investigations have revealed that it is not until about 4 years of age that children understand the effect that visual perception has on the mind. For example, Flavell et al. (1981) showed that, even though 3-year-olds could accurately judge whether another person could see an object, they did not understand that objects can give rise to different visual impressions depending on how they are seen. One interpretation of this kind of evidence is that before about 4 years of age children do not appreciate that visual perception can play a causal role in determining what others know.

Other studies confirm this general age transition in young children's understanding of the mental consequences of perception, both visual and otherwise. Mossler, Marvin, and Greenberg (1976) reported that 3-year-old children did not appear to understand that someone who had been told something possessed unique knowledge vis-à-vis someone who had not been told. Wimmer et al. (1988) provided evidence that, whereas 4–5-year-olds understood the role that both auditory and visual perceptual access plays in knowledge formation, 3-year-olds apparently did not. Gopnik and Graf (1988) demonstrated that 3-year-olds' difficulty in understanding perceptual access as a source of knowledge acquisition applies to their own states of knowledge as well. Additional studies have reported similar findings that converge to suggest that young 3-year-olds do not understand the role that perception plays in knowledge formation for either themselves or others (O'Neill & Gopnik, 1991; Perner & Ogden, 1988; Povinelli & deBlois, 1992; Ruffman & Olson, 1989; Wimmer et al., 1988).

This work is consistent with Flavell's earlier research that suggested that 3-year-olds lack an appreciation of the mental consequences of seeing and indicates that this difficulty may be a general one, extending to the other perceptual systems as well (although, for a discussion of potential differences in young children's understanding of different sensory modalities, see Flavell, Green, & Flavell, 1989). In short, prior to about 4 years of age, children may not understand the seeing-knowing relation. Other research, however, provides support for the view than many 3-year-olds do understand the seeing-knowing relation (Pillow, 1989; Pratt & Bryant, 1990; Woolley & Wellman, 1993; but see critiques by Povinelli & deBlois, 1992). Recent work by Lyon (1993) suggests that at least some of the disagreement may be resolved if the 3-year-old's theory of knowledge turns out to be one in which desire and knowledge are equated. His research shows that 3-year-olds attribute knowledge to others who show an interest in the situation over those who have had perceptual access but show less interest.

Do nonhuman primates also develop an understanding of the causal role that visual perception plays in knowledge acquisition? Early work by Menzel (1974) revealed that chimpanzees who were shown the location of a food reward by human experimenters were successful in attracting naive chimpanzees to follow them to the location of the reward. Later, the followers seemed to know which animal best knew the correct location of the reward and readily followed that animal to the spot. However, his studies were not explicitly designed to test the perception-knowledge hypothesis, and hence they contain a number of features that prevent a strong inference that the followers were making inferences about the connection between seeing and knowing. For example, the chimpanzee leaders provided strong behavioral cues to the followers such as tapping them, repeatedly glancing at them, and, "in the extreme case, screaming, grabbing a preferred companion, and dragging" (p. 115) them in the desired direction.

Experimental investigations of the seeing-knowing relation in nonhuman primates have been conducted with only three species: chimpanzees, rhesus macaques, and Japanese macaques (Cheney & Seyfarth, 1990a; Povinelli et al., 1990; Povinelli et al., 1991; Povinelli, Rulf, & Bierschwale, 1994; Premack, 1988). Both Premack (1988) and Povinelli et al. (1990) tested chimpanzees in an explicit effort to determine whether they understood the seeing-knowing relation. Premack (1988) reported that two of four 6–7-year-old subjects consistently selected a human trainer who had witnessed where food was being hidden as opposed to one who was behind a screen during the hiding procedure. Povinelli et al. (1990) allowed chimpanzees to choose between the pointing advice of two humans, a "knower," who had previously seen where food was hidden, and a "guesser," who had been out of the room during the hiding procedure. Three of the four subjects learned to respond preferentially to the knower and showed some evidence of transfer into a novel, but analogous, procedure. In contrast to the findings with chimpanzees, Cheney and Seyfarth (1990a), working with Japanese macaques, and Povinelli et al. (1991), working with rhesus macaques, reported an inability to demonstrate a comprehension of the seeing-knowing relation.

Despite the suggestive evidence to date, most researchers remain cautious about concluding that chimpanzees understand the seeing-knowing relation (or, indeed, anything else about mental states per se; see critical reviews by Cheney & Seyfarth, 1990b; Povinelli, 1993, 1996; Premack, 1988; Tomasello & Call, 1994; Whiten, 1993). This is due (in part) to the fact that previous research on chimpanzees' understanding of the seeing-knowing relation possesses serious methodological limitations. The report by Povinelli et al. (1990) suffers from the problem that in the initial phases of that experiment the subjects were administered weeks of repeated trials, raising the possibility that the subjects simply learned a discrimination. Povinelli et al. did employ a transfer test that was designed to determine whether the subjects were relying strictly on learned rules about observable events or whether they were making inferences about knowledge states resulting from differential visual access. Several of the subjects performed significantly above chance after 30 trials. However, their performances were not above chance during the first several trials of the transfer test, thus suggesting that they might have been the result of rapid learning of observable events (Povinelli, 1991, 1994a). In addition, both the Povinelli et al. and the Premack investigations suffer from the common limitation that the person who could see was differentially associated with a particular object or situation (Povinelli, 1991; Whiten, 1993). Finally, both studies suffer from employing relatively few manipulations involving visual deprivation, thus limiting the generality of the findings, even within the particular subjects that were studied.

ASSESSING YOUNG CHIMPANZEES' UNDERSTANDING OF VISION

The series of studies reported in this *Monograph* was designed to step back from the ambiguous results of chimpanzees' understanding of the seeing-knowing relation and attempt to establish whether they at least interpret vision as indicating the mental state of attention. The purpose of such a retreat is threefold.

First, as we noted in Chapter I, a comparative program designed to reconstruct the timing and order of the evolution of mental state attribution should examine both late and early emerging aspects of theory of mind in the human species. Such a research strategy can provide powerful leverage when making inferences about the validity of a given nonverbal measure of a specific form of mental state attribution (see Povinelli & deBlois, 1992; Premack & Dasser, 1991). Understanding vision as a subjective event that grounds organisms to the world appears to be one of the earliest-emerging understandings of mental phenomena that is constructed in human ontogeny (see Lempers et al., 1977).

The second reason for such a retreat is related to recent attempts to test for a coherence between the ability of individual members of a species to recognize themselves in mirrors and their tendencies to attribute mental states to others (Gallup, 1982, 1983, 1985). Although some work has been conducted in this area, inferences about phyletic differences in mental state attribution have been compromised by the complexity of the tasks involved. Povinelli et al. (1991), for example, have noted that the fact that rhesus monkeys fail to recognize themselves in mirrors and also fail to show evidence of understanding the seeing-knowing relation does not mean that they also lack a simpler mentalistic understanding of seeing, such as that related to level 1 visual perspective taking.

Finally, Premack (1988) has pointed out that what is needed is a wide range of studies that can ultimately provide us with an understanding of the chimpanzee's theory of seeing. From the perspective of the broader evolutionary framework outlined in Chapter I, constructing an accurate model of the development of the chimpanzee's theory of visual perception will provide us with a starting point for understanding when a mentalistic understanding of vision evolved.

III. UNDERSTANDING WHO CAN SEE YOU: PRELIMINARY INVESTIGATIONS

In order to assess whether young chimpanzees possess an understanding that visual perception connects organisms to the external world, we first constructed two theoretical frameworks (behaviorist vs. mentalistic)[1] to explain their behavior. These frameworks were used to generate predictions about how they ought to behave in circumstances where they were confronted with two individuals who differed in their visual connection to a given situation. The studies reported in this chapter and the next all utilized a common strategy: we presented young chimpanzees with two humans—one who could see them and one who could not—and then allowed them to use a species-typical behavioral gesture (a "begging" gesture; see Goodall, 1986) to request food from one of the humans. We explicitly chose a gesture that is frequently used by chimpanzees in general, and our subjects in particular, in order to maximize the ecological validity of the task. Indeed, we viewed this as an especially interesting gesture to examine because naive human observers invariably interpret the gesture as an act of intentional communication on the part of the chimpanzee. Thus, determining exactly what the subjects really understood about the gesture was of special interest to us.

The mentalistic framework assumes that chimpanzees understand that visual perception mentally connects them and others to the external world, and it thus predicts that they ought to gesture selectively toward others who can see their gesture. In contrast, because the behaviorist framework assumes that chimpanzees do not understand vision as an intentional pro-

[1] We use the terms *behaviorist* and *mentalistic* as equally pejorative or laudable descriptors for our frameworks. They are intended to convey the extreme ends on a spectrum of views about the factors controlling the behavior of organisms. In this *Monograph,* given that we are interested in testing for the presence of a mentalistic understanding of seeing, the null hypothesis is the extreme behavioristic position that the chimpanzees' behavior is completely governed by observable entities and events without recourse to reasoning about unobservable mediating mental states such as attention.

cess, it predicts that, if confronted with someone who can see them and someone else who cannot, the chimpanzees should gesture toward both with equal frequency. The behaviorist framework allows the subjects to process and use information about eyes and eye gaze and even to *learn* rules about at whom to gesture in such circumstances. However, unlike the mentalistic framework, it clearly predicts that, in the situation described above, chimpanzees should initially perform at random. In this fashion, we pitted the two frameworks against each other and evaluated their ability to predict correctly what the chimpanzees would do when confronted with a variety of situations that we designed to exemplify the concepts *seeing* versus *not seeing*.

EXPERIMENT 1

In the first experiment, we conducted a preliminary exploration of young chimpanzees' understanding of the subjective dimension of the visual attention of human experimenters. First, we trained the chimpanzees to use their begging gesture to "request" (hereafter no scare quotes) a food reward from a single experimenter. Next, we inserted occasional critical probe trials where we confronted them with two experimenters. On many of these probe trials, one of the experimenters could see their behavior, but the other could not. The mentalistic framework predicted that the subjects would gesture toward the experimenter who could see them. The behaviorist framework predicted chance-level performance on these probe trials because, although the framework acknowledges that the subjects can perceive the difference between the two experimenters, the significance of the differential visual access should be lost on them.

Method

Subjects and Housing

The subjects were six chimpanzees who ranged in age from 4 years, 4 months (4-4), to 5 years, 0 months (5-0), when their initial training began, from 4-8 to 5-4 when the testing for this experiment began, and from 5-4 to 6-0 when all the testing reported in this *Monograph* ended. Five of the subjects were female (Brandy, Jadine, Mindy, Megan, Kara), and one was male (Apollo). All the subjects had been born in captivity at the New Iberia Research Center and had either been transferred into a nursery at birth (Brandy, Kara, Mindy, Jadine) or been reared by their mother for about a year before being transferred into the general nursery (Megan, Apollo).

The nursery consisted of several large playpens with toys, blankets, and stuffed animals, sleeping cribs, and several human caretakers. At approximately 1 year of age, the subjects were transferred into a transition nursery with six additional peers in which they spent approximately half of each day in a outdoor play enclosure with access to an older group of chimpanzees through a wire fence. At approximately 2 years of age, the subjects were transferred to a large indoor-outdoor living area containing swings, barrels, tires, balls, and other hard plastic toys.

When they were between the ages of 2-9 and 3-7, five of the subjects became part of a long-term investigation of chimpanzee cognitive development. During the first year of the project, they continued to live with the other six members of their social group and were removed individually each day for testing sessions in a separate lab area, where they were tested in individual holding cages (see Povinelli et al., 1994). These five subjects were transferred to a specialized living and testing complex when they were between 4 and 5 years of age. The sixth subject (Jadine) was added to the long-term project approximately eight months later and had been a part of the subjects' original nursery group. A seventh subject (Candy), who had also been part of the original nursery group, was added to the long-term project group 4 months after Jadine. Candy thus began training on the current project several months later than the others and was not incorporated formally until sometime later (see Experiment 9 below). This subject was 4-6 when she began training, 4-10 when she began pilot testing, 4-11 when she began formal testing (see Experiment 9), and 5-2 when all the testing reported in this *Monograph* ended.

The layout of the specialized complex consisted of a series of three indoor-outdoor units that were completely interconnected by passageways that could be blocked off as needed. The animals had free range of this area (except the indoor testing room) at all times other than during testing periods. During these time periods, one animal at a time was transferred out of the group and tested. The animals were fed a standard diet of monkey chow, fruit, and vegetables twice daily, and this diet was supplemented by fruit, cookies, and (more rarely) candy that they received during testing.

The subjects were typically tested individually twice a day in a testing unit that consisted of a room with a shuttle door that connected it to one of the outdoor runs. The animals entered the testing unit through this door and were restricted to the rear third of the room by a clear Lexan partition and a ceiling caging panel. The subjects could be restrained to this area by closing the shuttle door to the outside and the door on the Lexan partition, or the doors could be left open as needed. A number of large holes (aligned horizontally at different heights) were cut into the Lexan. For the purposes of this experiment, the subjects were trained to use only two of the holes. These were the most extreme left and right holes, and they were at a height through which the subjects could comfortably reach. When these studies began, the animals were thoroughly habituated to manipulating a variety of apparatuses through all the holes as well as receiving food through them. The remainder of the room served as lab space for the experimenters and was devoid of any extraneous stimulation (only the experimenters, a table, data sheets, and a videocamera were present).

Both before and during the investigations described here, all the subjects participated in a number of cognitive studies involving the development of their capacity for self-recognition in mirrors (Povinelli, Rulf, Landau, & Bierschwale, 1993), their ability to understand the connection between visual perception and knowledge formation (Povinelli et al., 1994), their capacity to assess their own knowledge states, their reactions to accidental and intentional events (Povinelli, Perilloux, Reaux, & Bierschwale, 1995), as well as a variety of other capacities involving spatial memory, matching to sample, and the detection of agency on videotape stimuli. Although the subjects' primary social experiences were with other chimpanzees, all the subjects had extensive experience with human caretakers and, as part of their participation in the longitudinal research program, their trainer, research assistants, and students.

Procedure for Training the Subjects to Gesture

As part of an unrelated set of experiments, when the subjects were between the ages of 4-4 and 5-0, they were trained to gesture toward human experimenters (and novel actors) to bring them food rewards (Povinelli, Perilloux, Reaux, & Bierschwale, 1995). To accomplish this, we capitalized on their spontaneous begging gestures, which consisted of an outstretched arm with palm extended face up.

Training (as well as the later testing) involved several human participants. One of these was the animals' primary caregiver and trainer, and, throughout this *Monograph,* he is referred to as their *trainer.* Other human participants are designated as *experimenters.* The animals' trainer separated them from the group one at a time and ushered them into the indoor testing room. The procedure consisted of training the animals on a conditional discrimination task. The subjects watched as their trainer placed a large wooden screen in front of the holes in the Lexan partition, preventing the animals from reaching through. A small table was set up approximately 1.5 m in front center of the panel, and a small bowl containing small cookies and/or pieces of fruit was placed on its surface (see Fig. 3). Next, an experimenter positioned himself or herself on either the left or the right side of the table directly in front of one of the two relevant holes in the Lexan. Thus, after the trainer removed the screen, two options were available to the animals. They could gesture through either the hole in front of which the experimenter was standing or the hole on the side without the experimenter.

Initially, the subjects were verbally encouraged to reach through the correct hole in front of the experimenter. If they did so within about 30 sec after the trainer removed the screen, the experimenter immediately praised the animal, reached down to the table, picked up a small cookie or slice of fruit, and then stepped forward and handed it to the chimpanzee. The screen was then replaced, and the process was repeated. For this as well as all subsequent experiments, a response was considered to have been made if a subject's hand broke the plane of one of the holes in the Lexan; if subsequent responses occurred, they were disregarded. If the subject gestured through the correct hole, the animal was rewarded as described above; if the subject gestured through the incorrect hole, he or she was not rewarded, and a new trial began. Each animal was administered either 10 or 20 trials of this procedure a day until all were responding at 90% accuracy across 20 consecutive trials. The location of the experimenter (left or right position) on each trial was determined according to a predetermined quasi-randomized schedule.

After reaching the 90% criterion, a modification was introduced into the training procedure in which the shuttle door that connected the indoor testing unit to the outdoor run was left open and the trainer used the screen to usher the subject into the corresponding outside waiting area. To accomplish this, the trainer entered the testing unit, and, once the subject moved into the outdoor area, he positioned the screen so that it completely blocked the shuttle door opening, keeping the subject outside and preventing the subject from seeing into the testing unit. As before, an experimenter stood on either the right or the left position with the table with food still centered in the same position. Each trial began as the trainer removed the screen and immediately moved to the predetermined area outside the testing unit and visually focused on a neutral spot adjacent to the testing unit in order to minimize any potential cues that he might give the animals. With the shuttle door open, the subject was then free to enter the testing unit from the outside waiting area, approach the Lexan panel, and make a response by reaching through one of the holes.

As before, the task facing the animal was to gesture through either the left or the right hole, depending on the position of the experimenter. Again, correct responses (reaching through the hole that corresponded to the position of the experimenter) were reinforced and praised by the experimenter, whereas incorrect responses were not. No encouragement of any kind was offered prior to the subjects' responses. The subjects were trained until their responses averaged 95% correct or better across four consecutive 10-trial training sessions (38 of 40 or better). From this point forward, the experimenters fixed their gaze on a small target that was positioned exactly midway between the two response holes.

At this point, the subjects participated in an unrelated experiment in which they observed unfamiliar people engage in staged accidental or intentional events that resulted in the chimpanzees not receiving an expected food reward. The subjects were then allowed to choose between them. At the end of these tests, the subjects were tested for retention on the basic gesturing procedures described above. From here forward, we define this kind of trial in which the chimpanzees were required to gesture through either the right or the left hole to a single experimenter as a *standard trial.*

Pretest Procedure

On the day before the present series of experiments began, each subject was given a preference test that involved choosing between a block of wood and a food reward. The purpose of this test was to ensure that in later tests in which the experimenters would offer the subjects a choice between a block of wood and a food reward we could be certain that they had an unambiguous preference for the food.[2] The pretest session consisted of 10 trials. The subjects were kept in the outdoor run while two tables were set up in front of the right and left holes, respectively, within easy reach of the subjects. According to a predetermined counterbalanced schedule, a small block of wood was placed on one table, and a food reward was placed on the other. The subjects were released into the test room, and the first hole through which they reached was recorded. At this point, the subjects were all 4 months older than when the initial training began (the age range was from 4-8 to 5-4).

[2] For those who doubt the necessity of such a preference test, it is worth noting that our original design was one in which one experimenter offered a fresh, ripe banana and the other offered an extremely rotten banana. However, preference tests revealed that two of the six subjects consistently preferred the extremely rotten banana over the ripe one.

FIG. 3.—Experimental setting for training and testing the subjects. The sequence depicts a *standard trial* in which the subject is rewarded for gesturing through either the left or the right hole, depending on where the experimenter is standing. *a,* The trainer blocks shuttle door, thereby restricting the subject to an outdoor waiting area while an experimenter stands in front of the left response hole and fixes her gaze on the clear Lexan panel. *b–c,* The trial begins with the trainer removing the screen and exiting the testing unit, thereby allowing the subject to enter and respond. *d,* The subject responds correctly by gesturing through the hole directly in front of the experimenter. *e,* The subject is rewarded. For additional details, see the text.

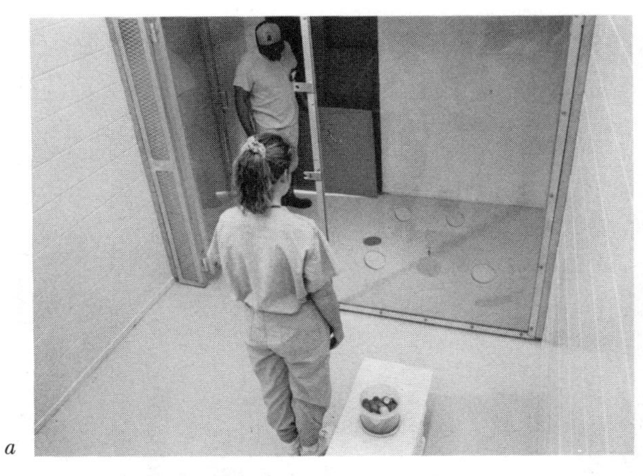
a

b

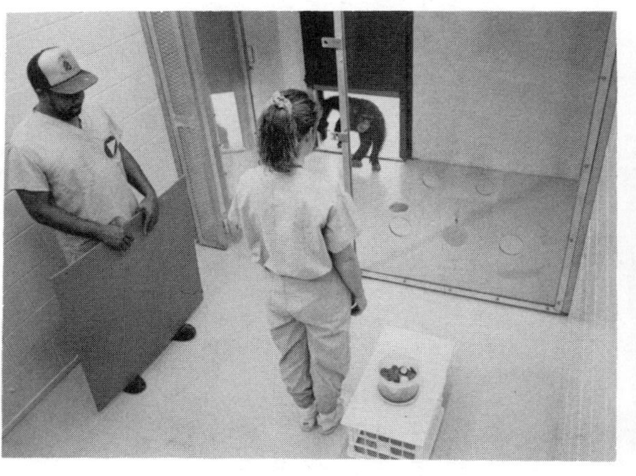
c

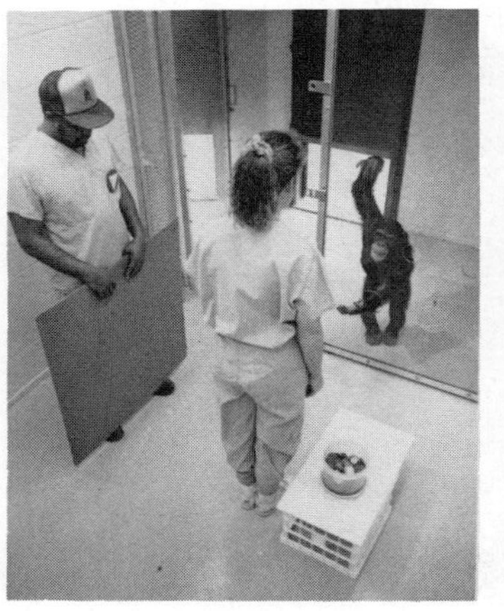
d

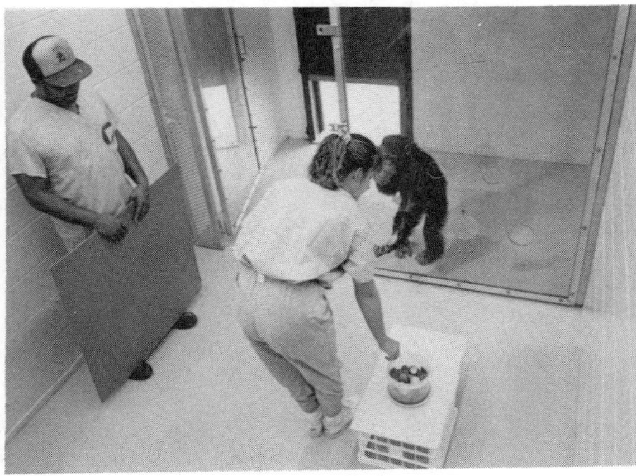
e

Testing Procedure

Note that up to this point the subjects' task had been to gesture through either the left or the right hole corresponding to the position of a single experimenter (standard trials). In contrast, the subjects were now tested by comparing their performances on different types of probe trials in which they were required to choose between two experimenters who stood in front of the Lexan partition. On some of these probe trials, only one of the experimenters could see them (thus, the mentalistic framework predicted that the subjects should gesture in front of that person). In these cases, we sought to determine whether the subjects would selectively gesture toward the one experimenter who could see them. If they did so, this experimenter picked up a food reward from the table between the two experimenters and rewarded the subject. If the subjects gestured in front of the experimenter who could not see them, they were not rewarded.

The subjects' performance on these probe trials was compared to their performance on other probe trials in which they were still required to choose between two experimenters but could do so on the basis of easily visible external signs (i.e., one experimenter offering a block of undesirable wood, the other offering a food reward). Again, if the subjects gestured toward the experimenter holding out the food, they were rewarded; if they gestured toward the one holding out the block of wood, they were not. These different kinds of probe trials involving a choice between two experimenters were embedded in a background of standard trials that, as defined above, involved only one experimenter. Below, we describe the various kinds of probe trials used in this study, and then we describe how they were administered to the subjects.

Procedure for Baseline Probe Trials

The purpose of baseline (A) probe trials was to provide a control on (1) the subjects' motivation to solve a problem involving two experimenters (as opposed to only one) and (2) the subjects' ability to compare the two experimenters and make a choice between them in a situation that did not involve making an attribution of seeing (or any other mental state). For these baseline (A) probe trials only, two identical small tables were set up in front of the two experimenters with no table in the middle position, as in the standard trials described earlier. One of the experimenters held out one-quarter of a banana or an apple or several cookies (depending on the food preference of individual subjects) toward the hole directly in front of him or her (Fig. 4). The table in front of this experimenter had a block of wood on it. The other experimenter held out a block of wood in an identical manner toward the hole in front of him or her. The table in front of this experimenter contained a clearly visible food reward of the same type and quantity as the other experimenter held (see Fig. 4).[3]

Procedure for Treatment Probe Trials

In each of the treatment probe trials, the subjects were faced with a choice between two experimenters, one whose vision was occluded and one whose vision was not occluded. One experimenter was positioned in front of the right hole and the other in front of the left hole as in the baseline probe trials. However, only one table was present, and it was positioned between them in the same location as on standard trials. The food rewards were placed by themselves in the center of the table (see Fig. 4). The treatments were of two logical types: two treatments (B and B′) using objects to obscure one of the experimenter's vision and two treatments (C and C′) in which one experimenter's vision was obscured in a more natural manner. Figure 4 depicts the contrasts present in each of the treatments. The probe trials in the blindfold treatment (B) consisted of both experimenters standing passively on the right or left, one with a blindfold covering his or her eyes and the other with an identical blindfold over his or her mouth. The probe trials in the bucket treatment (B′) consisted of both experimenters holding large white plastic buckets. One of the trainers held a bucket on his or her shoulder, the other over his or her head. For the probe trials in the back-versus-front condition (C), one experimenter faced the front of the Lexan partition as usual, whereas the other experimenter was positioned so that his or her back faced the partition. The probe trials in the hands-over-eyes treatment (C′) consisted of one experimenter with his or her palms completely obscuring his or her eyes, while the other experimenter stood in an identical manner except that his or her palms covered his or her ears instead. (The Appendix provides a detailed list and description of all these treatments, along with ones used in subsequent experiments.)

For all testing trials, including all standard, baseline, and treatment probe trials, the experimenters fixed their gaze on the target midway between the two response holes. We choose this instantiation of "seeing" over

[3] An exploratory baseline test of two sessions (each session = 10 trials) was initially conducted in which one experimenter stood on the right and the other on the left on every trial. One held out an arm offering the subject a block of wood, whereas the other held out an arm offering a food reward. Although the subjects performed nearly flawlessly (only a single error in the whole group), we were concerned that they might be responding merely to the presence of food and not to the offering gesture per se. We therefore designed the baseline (A) using the tables in front of the experimenters, an arrangement that was used for the remainder of the experiment.

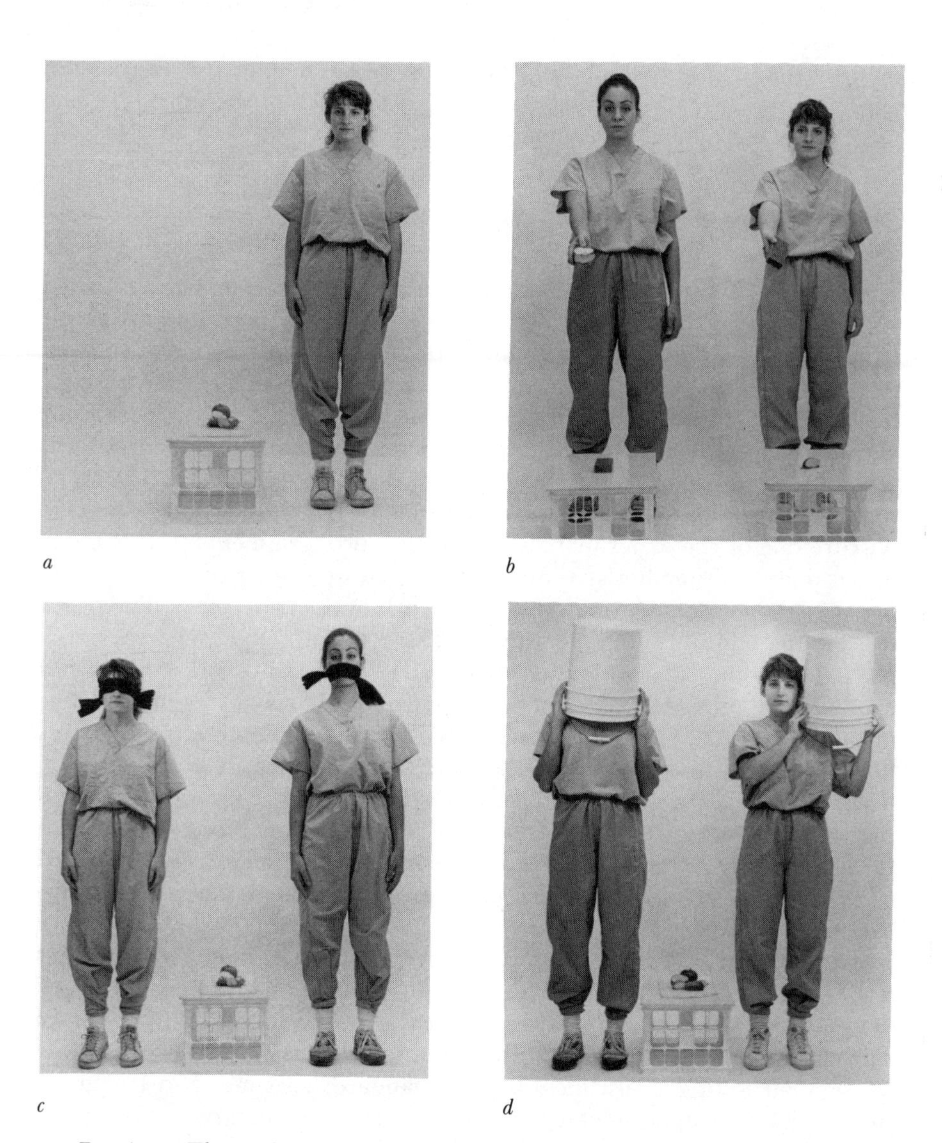

Fig. 4.—*a,* The configuration for standard trials that surrounded all probe trials (see also Fig. 3 above). *b,* The configuration for baseline probe trials (A) in which subjects chose between one experimenter offering food (the experimenter on the left) and another offering a block of wood (the experimenter on the right). (Note the position of the tables that contain the opposite stimuli of what each experimenter is offering.) *c,* The stimulus configuration for blindfold probe trials (B). *d,* The stimulus configuration for bucket probe trials (B′).

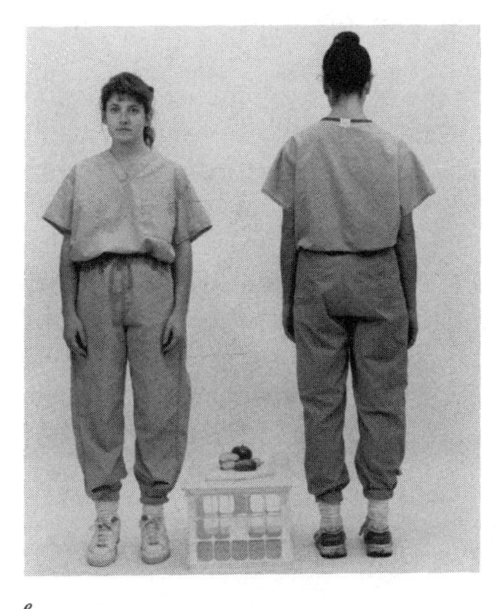

e *f*

FIG. 4 (*Continued*).—*e*, The stimulus configuration for back-versus-front probe trials (C). *f*, The stimulus configuration for hands-over-eyes probe trials (C′).

the option of having the experimenters attempt to make direct eye contact with the subjects as they entered the testing unit for several reasons. First, if (as seems likely with chimpanzees) direct eye contact can be an attractive stimulus, then this might draw them to that experimenter even in the absence of an understanding of the attentional aspect of seeing. In addition, we reasoned that, if chimpanzees truly understood that seeing connects someone to the external world, it should not matter whether the subjects see the experimenter looking at them directly or where they are about to respond. This distinction can be fruitfully thought of in the context of Brothers and Ring's (1992) distinction between "hot" and "cold" aspects of theory of mind. They note that social stimuli that have an emotional ("hot") valence for organisms may be conceptually distinguished from more cognitive ("cold"), intentionally based interpretations of social action. In this case, we see the potential attraction to direct eye contact as raising the operation of "hot" aspects of theory of mind, not the cognitive ones in which we are primarily interested here. More generally, the level 1 perspective-taking studies reviewed in Chapter II reveal that children as young as 2½ years interpret seeing as a subjective event connecting the observer to the observed, regardless of whether the thing observed happens to be the subject or another object (e.g., Lempers et al., 1977). We thus decided to create a situation where one of the experimenters was connected to the scene (looking at the partition where the subject would respond) and to defer to future research the more theoretically complicated issue of an attraction to direct eye contact.

Blindfolds and buckets were chosen as the objects to obscure the vision of the experimenters, in part, because all the subjects had extensive previous experience with these items and/or similar items since birth (e.g., other buckets, burlap sacks, blankets, paper bags, large pieces of opaque plastic, PVC pipes and caps, opaque plastic and wooden boxes, tires, and barrels). In addition, the exact items used in the tests were introduced into the subjects' enclosure (and left there) the day before the respective tests. The subjects were repeatedly observed using them in ways that resulted in visual occlusion, such as placing buckets or burlap sacks over their heads. Indeed, in some cases, we observed the subjects placing their hands over their eyes and walking about the cages.

Testing Design

The subjects were tested using a design in which the four types of treatment probe trials (B, B′, C, C′) were presented to the subjects and were contrasted to baseline (A) probe trials. Sessions containing baseline or treatment probe trials are referred to as *baseline sessions* and *treatment sessions,* respectively. Within both session types, Trials 1–4 and 6–9 were standard trials (identical to the training procedures in which one experimenter was standing in front of either the right or the left hole with the food table in the middle). Trials 5 and 10 were designated as the critical probe trials in which both the right and the left positions were occupied by the experimenters. In these probe trials, the position of the experimenters and the food table(s) differed across the probe trial as described above. Both of the probe trials in a given session were of the same type (i.e., they were both baseline trials, or they were both the same type of treatment trials). For purposes of illustration, Table 1 depicts the structure of a typical baseline phase and treatment phase (each comprising two 10-trial sessions).

The baseline and treatment sessions were further grouped in phases where a baseline or treatment phase consisted of two consecutive baseline or treatment sessions. Each subject was administered the following specific order of baseline and treatment phases (consisting as described above of two sessions per phase): A-B-B′-A-C-A-C-C′-A. Thus, in each baseline and treatment phase, each subject received four probe trials (two per session) in which they chose between two experimenters. For example, in the first baseline phase (A), each animal received two probe trials during Session 1 and two probe trials during Session 2 for a total of four baseline probe trials. This provided us with two types of controls on the subjects' performance

TABLE 1

STRUCTURE OF BASELINE AND TREATMENT PHASES, EXPERIMENT 1

Trial	Baseline Phase, Sessions 1 & 2	Treatment Phase, Sessions 3 & 4	Baseline Phase, Sessions 5 & 6	. . .[a]
1–4	Standard trials (1 experimenter; hands at side)	Standard trials (1 experimenter; hands at side)	Standard trials (1 experimenter; hands at side)	. . .
5.	Baseline probe (A) (2 experimenters; E1: offers food, E2: offers block)	Treatment probe (B) (2 experimenters; E1: eyes blindfolded E2: mouth blindfolded)	Baseline probe (A) (2 experimenters; E1: offers food, E2: offers block)	. . .
6–9	Standard trials (as above)	Standard trials (as above)	Standard trials (as above)	. . .
10.	Baseline probe (A) (as above)	Treatment probe (B) (as above)	Baseline probe (A) (as above)	. . .

[a] The full design continues as is described in the text.

on treatment probe trials. Each treatment (and baseline) probe trial was surrounded by standard trials (providing a general motivation control), and each phase containing treatment probe trials was bracketed by phases containing baseline probe trials (providing a temporal control on the subjects' motivation to solve the problem when two experimenters composed the stimulus configuration but no inference about the mental state of attention was required to solve it). (For an excellent discussion of the logic of small-*N* experiments, see Robinson & Foster, 1979.)

The sessions were executed in a fashion similar to the training sessions. The chimpanzees' trainer ushered the subject into the outdoor run at the start of each trial and blocked the entrance to the test room with a wooden screen. Once the experimenter(s) were in place, the trainer removed the screen and exited the testing unit. The subject then entered and gestured through one of the two holes. Occasionally (two of the 216 probe trials in this experiment), a subject entered the room but did not respond within 30 sec. On these occasions, the trainer had been instructed to usher the subject out of the testing unit and repeat the trial until the subject responded.

A series of randomized schedules were constructed for the standard trials in such a way that each side (right and left) was correct equally often and each experimenter appeared equally often on both sides across sessions. For the critical probe trials, within each session the experimenters were positioned on the same side for each of the two trials, but each experimenter executed each alternative once (i.e., in the baseline sessions, if an experimenter held out the food on the first probe trial, he or she held out the block on the second). The order in which the experimenters were assigned their roles between sessions was assigned randomly and then counterbalanced across subjects so that half the subjects received one test order and the other half received the opposite order. All trials were videotaped.

Data Analysis

Each subject's responses on the two probe trials in each session were averaged, resulting in a score of either 0%, 50%, or 100% correct. Next, the six subjects' individual scores were averaged to yield an average percentage correct for the group on the probe trials for each session, and these scores were plotted across sessions. The subjects' responses were analyzed statistically by a one-way repeated-measures ANOVA comparing all subjects' average scores for each phase to each other. For this analysis, each subject was assigned a score for each phase (the two sessions of each treatment type combined) consisting of the percentage correct of four probe trials. Thus, each animal received a score of either 0%, 25%, 50%, 75%, or 100% correct for each phase.

Finally, videotapes of all probe trials (except one that was missing because of a taping error) were observed independently by two students to determine whether the subjects hesitated before making a choice. The purpose of coding these hesitations was to provide us with at least a crude measure of changes in the cognitive-emotional state of the subjects from one trial to the next. As we will explain, such changes in the subjects' states appeared to be tied to particular treatments.

In order to measure the hesitations, we instructed the students to observe the videotaped recording of each probe trial and to record (on preconstructed scoring sheets) all trials in which the subjects appeared to "hesitate" or "change their mind" by either pausing or detouring away from one side to the other before gesturing through one of the holes after the trainer had removed the screen and the subjects had entered the test room. There were two probe trials in which the subjects did not respond (i.e., they entered the test room but did not make a choice, or they sat in the doorway for more than 30 sec without making a choice). The observers were instructed ahead of time to score these trials as containing a hesitation, regardless of what the subjects did when the trial was repeated. An interobserver reliability index was calculated by the following formula: percentage agreement = total number of agreements ÷ total opportunities for agreement. For each animal, the two observers' scores were averaged to yield a percentage of probe trials during which the subject showed hesitations in each treat-

TABLE 2

RESULTS OF TRAINING CHIMPANZEE SUBJECTS TO GESTURE TO HUMAN EXPERIMENTERS, EXPERIMENT 1

Subject	Original Training: Trials to Criterion[a]	Retention[b]	
		Retention Test (% Correct)	Trials to Criterion[c]
Kara	250	100	50
Jadine	750	100	50
Brandy	410	100	90
Megan	430	80	60
Mindy	310	100	50
Apollo	560	100	80
Candy[d]	640	...	...
M	479	97	63
SD	180	8	18

[a] Criterion was a minimum of 38 of 40 consecutive trials correct.

[b] Retention test occurred roughly 2 months after the original training was completed and consisted of 10 trials.

[c] Retention criterion was set at 50 consecutive correct responses.

[d] Candy was trained after the other subjects and was not used in Experiments 1–8 (see Experiment 9).

ment phase. From this data, an overall group mean percentage of trials containing hesitations was calculated for each phase.

Results

Training to Gesture

All subjects were performing at above-chance levels within the first few training sessions. However, they took considerably longer to meet the stringent training criterion established ahead of time. They required between 250 and 750 trials to meet the criterion of 38 of 40 trials correct or better at gesturing through the hole in front of the trainer (see Table 2). After the subjects had completed the unrelated experiments involving accidental and intentional behavior, they were tested for their retention of the gesturing response. Five of the six subjects were 100% correct in their first retention test session (10 trials), and all required fewer than 10 sessions (10 trials each) to reach the retraining criterion of 50 consecutive trials correct (see Table 2).

Pretest and Testing

On the pretest involving a choice between reaching out to obtain a food reward or a block of wood, each of the subjects selected the food item on all 10 trials.

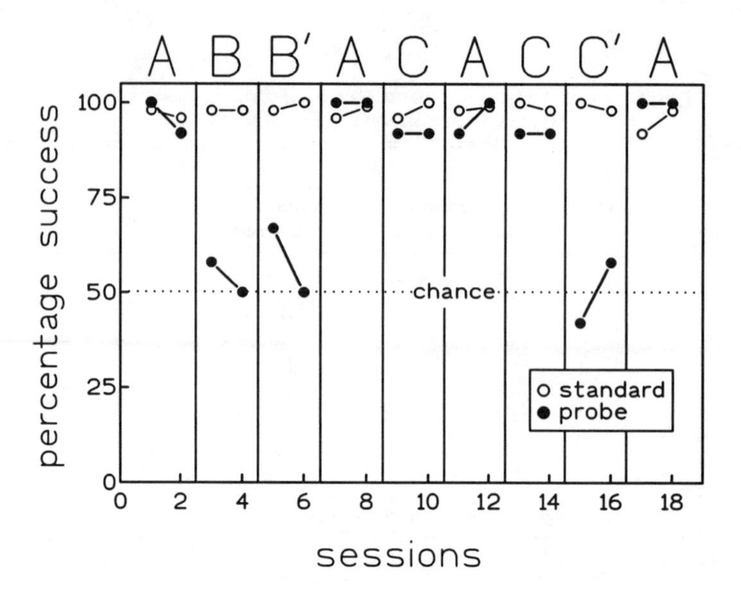

FIG. 5.—Overall results by session, Experiment 1. Each data point represents the grand mean of two trials per subject (N = 6 subjects). For details of the treatments, see the text. A = baseline; B = blindfolds; B′ = buckets; C = back-versus-front; C′ = hands-over-eyes.

Figure 5 provides a session-by-session summary of the subjects' performance across the nine phases of the experiment. The data points represented by the open circles reveal clearly that the subjects performed at ceiling levels (over 98% correct) on the standard trials (one experimenter present, no offering of food) across all nine phases of the experiment. This demonstrates that the subjects were attending and highly motivated to respond even when the experimenter was not holding out an arm and offering food. With respect to the subjects' performance on the probe trials, in the first two sessions of baseline (A), the subjects performed at near-perfect levels with only one subject making a single error. Because there was food on both the right and the left sides (either in the experimenter's hand or on the table), this demonstrated that the subjects were focusing their attention, not on the mere presence of food, but more specifically on what was in the experimenters' hands.

Objects as Causes of Visual Occlusion

In contrast to their strong preference for the experimenter who offered a food reward on the probe trials of the first baseline sessions, the subjects' performances on the probe trials dropped dramatically in both sessions with the blindfold treatment (B). Not only were the subjects' scores low in the first session, but they declined further in the second session. The same general pattern was true of the two sessions involving one experimenter with his or her entire head obscured by a bucket and another experimenter holding a bucket on her or his shoulder (see Fig. 5, Phase B′). Two aspects of the subjects' performance indicate that their random responses on the B and B′ probe trials can be attributed to the nature of the treatments and not other factors (such as a loss of motivation). First, their probe trial performance returned to perfection in the following two sessions involving the baseline probe trials in which the choice between experimenters involved an easily visible discrimination as opposed to an inference about the importance of the trainers' differential visual access. Second, the subjects continued to respond at or near ceiling levels on the standard gesturing trials within all sessions of B and B′ (see Fig. 5). Both of these results are critical to a strong inference about the cause of the effect witnessed on the treatment probe trials themselves.

"Naturalistic" Causes of Visual Occlusion

In striking contrast to the results from the two treatments involving objects (blindfolds, buckets), five of the six chimpanzees performed perfectly on the four probe trials involving one experimenter who faced the front of the testing unit and the other who faced away (Fig. 5, Phase C). One of the subjects, Mindy, responded at chance (50% in both sessions). Because we were struck by the robustness of this effect in contrast to the chimpanzees' performance when blindfolds and buckets were used as sources of visual occlusion, we attempted to replicate it. First, we returned the subjects to baseline (A) for two sessions, during which they made only one error on the probe trials. Next, we applied the back-versus-front treatment, and as a group the subjects made only two errors in the two sessions. Finally, all subjects received two sessions involving probe trials of the second naturalistic treatment (C′) in which the subjects chose between an experimenter covering his or her ears and an experimenter completely covering her or his eyes with the palms of his or her hands. In contrast to the results of the first naturalistic treatment, the subjects responded at chance levels during both sessions (see Fig. 5, Phase C′). That this drop in performance was the result of the treatment is apparent both by the subjects' performance on the final two sessions of baseline and by their continued ceiling performance on the standard gesturing trials (Fig. 5).

Separate one-sample t tests (two tailed) were performed on the group's average percentage correct scores for each of the treatment phases (B, B′, C, C′) using 50% as a hypothetical population mean (the percentage correct expected if drawn from a population responding at chance). The results confirmed that, in the blindfold, bucket, and hands-over-eyes phases, the subjects' performance did not differ from chance (50%) ($p > .36$ in all cases). In contrast, the subjects' performance in the first and second back-versus-front phases did differ significantly from chance ($t[5] = 5.0$, $p = .004$, and $t[5] = 7.9$, $p = .0005$, respectively). That this disposition was present immediately is further confirmed by the fact that, on their very first trial of this treatment, five of six subjects gestured toward the person facing forward and, on their second trial, all six animals did so. Finally, a one-way repeated-measures ANOVA revealed an overall effect of phase ($F[8, 40] = 11.23$, $p < .0001$). Tukey-Kramer multiple comparison tests revealed that none of the baseline treatments (A) differed from each other, nor did they differ from either of the back-versus-front treatments (C), which did not differ from each other. In contrast, all pairwise comparisons of baseline and back-versus-front treatments to the three other treatments involving visual occlusion (B, B′, C′) yielded significant differences in performance ($p < .05$ or smaller). The subjects' performances under these latter treatments (B, B′, C′) did not differ from each other. The pattern of these results confirms the overall impression from Figure 5 that the subjects responded significantly differently to the back-versus-front treatment as compared to the three other visual occlusion treatments.

Hesitations on Probe Trials

The two raters agreed as to whether the subject hesitated before responding on 91% of the 215 probe trials that they scored. The average percentage of probe trials in which the subjects displayed a hesitation before responding ranged from 0% to 42% across the nine phases. In order to determine whether the subjects hesitated more in some phases than in others, the hesitation data were subjected to a one-way repeated-measures ANOVA. The results revealed an overall significant effect of phase ($F[8, 40] = 4.12$, $p = .0012$). Post hoc Tukey-Kramer tests revealed that the overall effect was due to the subjects hesitating more in the bucket phase than in the second, third, and final baseline phases and the first back-versus-front phase ($p < .05$ or smaller in all cases). None of the other phases significantly differed from each other.

Discussion

A comparison of the results of each phase indicates that in three of four treatments the subjects responded in a manner suggesting that they did not take into account whether the person they gestured toward could see them perform the action. These results are consistent with the predictions generated by the behaviorist framework. That is, when confronted with two experimenters on the blindfold, bucket, and hands-over-eyes probe trials, the subjects did not selectively gesture toward the experimenter who could see them. In contrast, in the back-versus-front treatment, the subjects did respond in accordance with the mentalistic hypothesis by consistently gesturing toward the person who was facing them (for related data on two orangutans' ability to distinguish between "front versus back," see Call & Tomasello, 1994). It is important to note that all these findings reflect stable dispositions on the part of the subjects that were not learned as the result of repeated exposure to this kind of treatment probe trial. Thus, the subjects did not learn to respond correctly in the baseline condition or the back-versus-front treatment; rather, they performed "correctly" from Trial 1 forward. Likewise, the subjects showed no evidence of improving across the four trials of the other treatments that they were administered (B, B′, C′).

In considering the potential contradiction inherent in the finding that the chimpanzees immediately showed a preference for the experimenter facing forward but showed no preference for the experimenter who could see them in the other treatments, it is important carefully to consider the two general frameworks that we initially outlined. First, the differential performance in the back-versus-front treatment as compared to the other treatments involving visual occlusion could be interpreted as indicating that in the most natural of the conditions (i.e., the situation that most closely approximates experiences that the subjects had on a daily basis) the chimpanzees performed exactly as the mentalistic framework predicted. This could mean that, despite their random performance in the other treatments, the subjects did, in fact, understand the necessity of the experimenters being able to see their gestures in order for them to act on them but did not demonstrate this comprehension for procedural reasons.

Corollary arguments could be constructed to bolster this view by explaining the results of the other conditions as artifacts of the design. For instance, perhaps the subjects' believed that the experimenters were peeking through their hands or the blindfolds (although the case is more difficult to make for the bucket condition). Likewise, perhaps the subjects were emotionally aroused at the presence of the blindfolds or buckets, and this disrupted their performance. Some support for this hypothesis comes from the fact that the subjects hesitated most before making their choices in both these conditions (on 21% of trials involving blindfolds and 42% of trials involving buckets). The general thrust of these arguments is that in the most "natural" treatment the chimpanzees did display an understanding of the importance of seeing in the context of a communicative exchange. Some data that bear on this question have been recently reported by Tomasello, Call, Nagell, Olguin, and Carpenter (1994), who found that, in a naturalistic study, chimpanzees restricted visually based communicative gestures to situations in which the recipient was visually oriented toward them.

Alternatively, the typical failure of the subjects to take into account the visual access of the experimenters can be interpreted as support for the behaviorist view of their behavior. From this perspective, the animals' selection of the experimenter facing them in the back-versus-front treatment could be interpreted as a consequence of their reinforcement history in the training phase. Prior to the introduction of the back-versus-front treatment, the subjects had never been rewarded for gesturing toward the stimulus of the back of a person. In the hundreds of training trials preceding their first back-versus-front probe trial, the correct stimulus always involved the front of a person (front torso visible, head and feet pointing forward, position of thumbs and hands, etc.). This is also true of the probe trials leading up to this point in conditions A, B, and B′. Thus, when confronted with the back of an experimenter and the front of an experimenter, the subjects may simply have been attracted to the stimulus that most closely approximated the one previously rewarded. From this perspective, the subjects' responses may have had nothing to do with the visual deprivation of the experimenter facing away but may instead have been governed by the fact that the back of an experimenter was not sufficiently similar to the consistently rewarded stimulus to result in generalization (and hence responses toward it).

Several points about this explanation should be carefully noted. First, it does not specify the exact nature of the stimulus properties necessary to produce generalization. For instance, just as it was true that in every training trial the experimenters were facing forward, so too were their eyes and faces visible. Yet the subjects' performances dropped to chance levels when both these properties (eyes and face) were missing from one of the experimenters in the bucket condition. Thus, in that case, the subjects did not respond to the person who most closely matched the previously rewarded stimulus (the experimenter holding the bucket on his or her shoulder). Of course, it could be maintained that the general "frontalness" of the experimenter with the bucket completely over her or his head was still sufficiently similar to the original stimulus to result in the subjects' choosing that experimenter on half the trials.

EXPERIMENT 2

In the second experiment, we addressed two issues related to the subjects' performances in Experiment 1. First, we sought to replicate the major findings of that experiment using a within-session design of mixed probe trial types. The purpose of the replication was to investigate the robustness of the effects observed in the first experiment. The use of a mixed within-session design was an attempt to accelerate the process of data collection in order to reduce the possibility that the subjects would become less responsive in the test situation across time. The second issue that we addressed concerned the possibility that the subjects were using cues other than those we were manipulating as a basis for making their choices. Although we had no reason to suspect that this was the case (especially given that the subjects performed randomly on most of the critical probe trials), we nonetheless sought to rule out the possibility by covertly creating a bogus condition in which we intentionally misled the trainer and experimenters about the real nature of the test that they were administering to the subjects.

Method

Subjects

The subjects were the same six chimpanzees that participated in Experiment 1. The current experiment began 2 days after the completion of Experiment 1.

Procedure

The structure of the test sessions was identical to that used in the previous study except that a within-session design was employed in which four probe trials were interspersed within 20 trial sessions. Thus, Trials 5, 10, 15, and 20 were designated as target probe trials, with each being the vehicle for one of four treatments. Three were the same as in Experiment 1: back-versus-front, hands-over-eyes, and blindfold treatments were presented exactly as described in Experiment 1. A fourth control treatment (CH)[4] was created to determine whether the chimpanzees could detect and utilize unmanipulated cues from the trainer and the experimenters about the correct selection. We constructed a detailed (bogus) hypothesis and then disseminated it at a weekly lab meeting to the students and staff. The hypothesis

[4] *CH* is an acronym for Clever Hans, the horse who was exhibited by his trainer around the turn of the century and billed as performing a number of apparently "intelligent" acts, including simple addition. However, on careful examination of the horse and his trainer, a psychologist named Pfungst was able to demonstrate that the horse was utilizing physical cues (body inclination, eyebrow orientation) provided (apparently unconsciously) by his trainer (see Umiker-Sebeok & Sebeok, 1980).

centered around the fact that we purportedly suspected that the chimpanzees were carefully monitoring the position of the experimenters' hands because their hands were most closely associated with delivery of the food. We explained that we believed that the chimpanzees would select the person whose hands were most visible.

In order to set up the CH treatment, we informed the trainer and student experimenters that a fourth condition was being investigated in which both experimenters would stare forward as usual with their hands at their sides, except that one of the experimenters would have his or her hands resting on his or her thighs and the other would have his or her hands slightly obscured behind his or her legs. They were informed that the person whose hands were visible to the subjects would be correct and that, if the chimpanzees gestured toward this experimenter, they were to be rewarded. However, if the subjects gestured toward the experimenter whose hands were slightly obscured, no reward was to be delivered, and the trainer was to usher the subject out of the test room in preparation for the next trial. Every effort was made to convince the trainer and the student experimenters that the chimpanzees would probably be able to use this cue to select the correct experimenter. That the deception was effective was supported by subsequent debriefing sessions with the experimenters and trainers.

The subjects were each administered four 20-trial sessions (each session = 16 standard trials plus four mixed probe trials). Thus, by the end of the experiment, each subject received four trials of each of the four treatments. Six test schedules were constructed, and each subject was assigned four of the six so that for each subject the major variables (order of treatments, position of correct answer, position–trial type interaction) were counterbalanced within the constraints of the sample size. Each of the four probe trial types was assigned once to each of the four target trial positions (5, 10, 15, 20) within each test session, and across sessions each treatment occurred in each position an approximately equal number of times. Across the six test schedules, the right and left positions and each experimenter were correct an approximately equal number of times for each probe trial type. Each of the intervening standard trials was counterbalanced in terms of experimenters and positions.

The main results were summarized by calculating the mean performance of each animal on the four trials they received of each probe trial type (either 0%, 25%, 50%, 75%, or 100% correct) and then averaging the scores across subjects. A one-way repeated-measures ANOVA was used to determine whether there was an effect of treatment. The videotapes of all probe trials were viewed by two student assistants, who coded them for hesitations as described in Experiment 1.

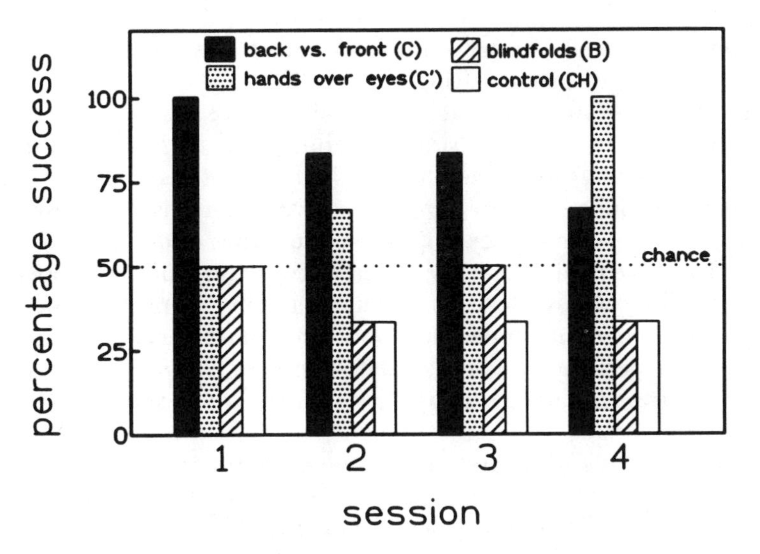

FIG. 6.—Overall percentage success by session, Experiment 2. For details of the treatments, see the text.

Results and Discussion

The results of this experiment are presented by session in Figure 6. Separate one-sample t tests (two tailed, hypothetical mean = 50%) for each condition averaged across all sessions indicated that the subjects performed at above-chance level in both the back-versus-front ($t[5] = 4.00$, $p = .01$) and the hands-over-eyes ($t[5] = 3.16$, $p = .02$) conditions. The group performance did not differ significantly from chance in either the blindfold or the control conditions. A repeated-measures ANOVA revealed an overall effect of phase ($F[3, 15] = 6.87$, $p < .005$), and Tukey-Kramer post hoc tests indicated that the animals' performances were significantly better in the back-versus-front treatment than in both the blindfold condition ($p < .01$) and the control condition ($p < .05$). No other conditions differed from each other.

In general, the results depicted in Figure 6 confirm our findings from Experiment 1 that the subjects responded at chance in the blindfold (B) treatment but not in the back-versus-front (C) treatment. They also responded at chance in the control (CH) treatment. This latter result strongly suggests that the chimpanzees were not using some kind of unconscious cues from either their trainer or the experimenters. Indeed, given that their performance was at chance levels in most treatment conditions in both

Experiment 1 as well as most of this experiment, it is difficult to make the case that the subjects were relying on such cues.

The subjects' intermediate and above-chance performance in the hands-over-eyes treatment warrants careful attention. A closer inspection of the session-by-session results reveals that the results of the first three sessions were consistent with Experiment 1: the animals were responding at chance in all treatments except back-versus-front (see Fig. 6). However, the results of the final session are different from those of both Experiment 1 and the previous three sessions. Notably, the subjects' performance in the back-versus-front treatment declined (only four of six animals were correct), whereas their performance in the hands-over-eyes treatment improved (six of six animals were correct). Thus, they appeared to be showing a learning effect across sessions.

The two raters who coded the probe trials for hesitations agreed on 92% of the 95 probe trials that they scored. The average percentage of trials in which the subjects displayed hesitations before making a choice ranged from a low of 4% in the second baseline condition to 56% in the blindfold condition. The results of a one-way repeated-measures ANOVA of the hesitation data revealed an overall significant effect ($F[3, 15] = 15.52$, $p = .0001$), and Tukey-Kramer post hoc tests indicated that the effect was due to the subjects hesitating significantly more often on the blindfold probe trials than on the other three probe trial types ($p < .001$ in all three cases).

The decline in performance across test sessions in a treatment previously understood by the subjects (back-versus-front) and the improvement in a treatment previously not understood suggested one of two possibilities. First, it was possible that the differential reward received by the subjects had trained them to pick the person whose eyes were not obscured by his or her hands. This explanation would seem to suggest that the subjects should also have learned a comparable rule concerning the blindfolds because at that point they had received the exact same number of probe trials using each treatment ($N = 8$). However, they were still performing at chance on the blindfold probe trials. Another possibility (which we considered more likely) was that across the sessions the subjects may have learned to anticipate being incorrect half the time on three of the four probe trials in each session (blindfolds, control, hands-over-eyes). Thus, by Session 4, they may have been responding randomly, and their poor performance in a treatment in which they were previously successful (back-versus-front) and their improvement in one in which they were not previously successful (hands-over-eyes) reflect chance fluctuations. Mixing probe trial types within sessions appeared to interfere with their previously stable performance on a treatment on which they previously appeared to have a strong disposition.

EXPERIMENT 3

In the next experiment, we returned to our initial design in order to answer several questions related to the animals' performance in the first two experiments. First, we wished to determine whether we could replicate the effect obtained in the final hands-over-eyes session of Experiment 2; if we could not, it would rule out the learning hypothesis and implicate the interference hypothesis. Likewise, we sought to determine whether the subjects no longer had a disposition for gesturing in front of the experimenter facing forward in the back-versus-front treatment. Finally, we attempted to test the hypothesis that the subjects' rather robust tendency to select the trainer facing forward was due to the fact that the experimenter could see them. With respect to this last hypothesis, we reasoned that the mentalistic framework predicted that, if the chimpanzees were confronted with two experimenters, both with their backs facing the testing unit, but one looking over his or her shoulder toward the subject, the chimpanzees would gesture toward the person who could see the gesture. In contrast, the behaviorist framework predicted random responding if the stimulus governing the chimpanzees' choices was simply the front of a person.

Method

Subjects

The subjects were the same six animals used in the previous investigations. The subjects began this experiment 2 days after completing Experiment 2.

Procedure

The procedure was identical in general structure to that described in Experiment 1. In other words, each subject was administered 10 trials per session with Trials 1–4 and 6–9 as standard gesturing trials with one experimenter on either the right or the left positions. Trials 5 and 10 were designated as probe trials with two experimenters present. As in Experiment 1 (but not Experiment 2), both probe trials within a session were of the same type.

The animals were tested using an A-C′-A-C-C″-A design, where A, C′, and C corresponded to the same treatments used in Experiments 1 and 2 (baseline, hands-over-eyes, and back-versus-front, respectively). The new condition (C″) consisted of probe trials where the two experimenters both presented their backs to the front of the testing unit but one experimenter

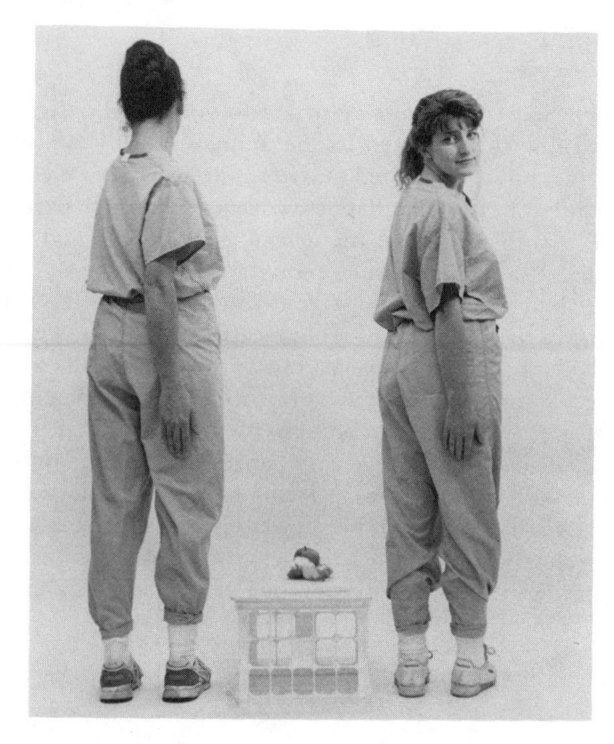

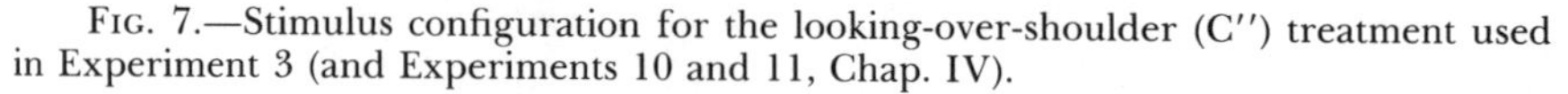

FIG. 7.—Stimulus configuration for the looking-over-shoulder (C″) treatment used in Experiment 3 (and Experiments 10 and 11, Chap. IV).

looked over his or her right shoulder back toward the visual fixation target on the clear Lexan partition. The other experimenter angled his or her shoulders slightly to match the experimenter who was looking over his or her shoulder but looked directly at the wall opposite the testing unit (see Fig. 7). On each probe trial, before the subject was let into the room, the trainer carefully checked the orientation of both experimenters and adjusted them as needed so that they were as identical as possible in overall body orientation, except for the aspect of looking over the shoulder. The data were collected and analyzed in the same fashion as in the previous experiments.

Results and Discussion

The overall results are presented in Figure 8. As in Experiment 1, the subjects performed at near-perfect levels on the standard trials (over 98% correct), thus providing a within-session control on their motivation to respond. In contrast, separate two-tailed one-sample *t* tests (hypothetical mean

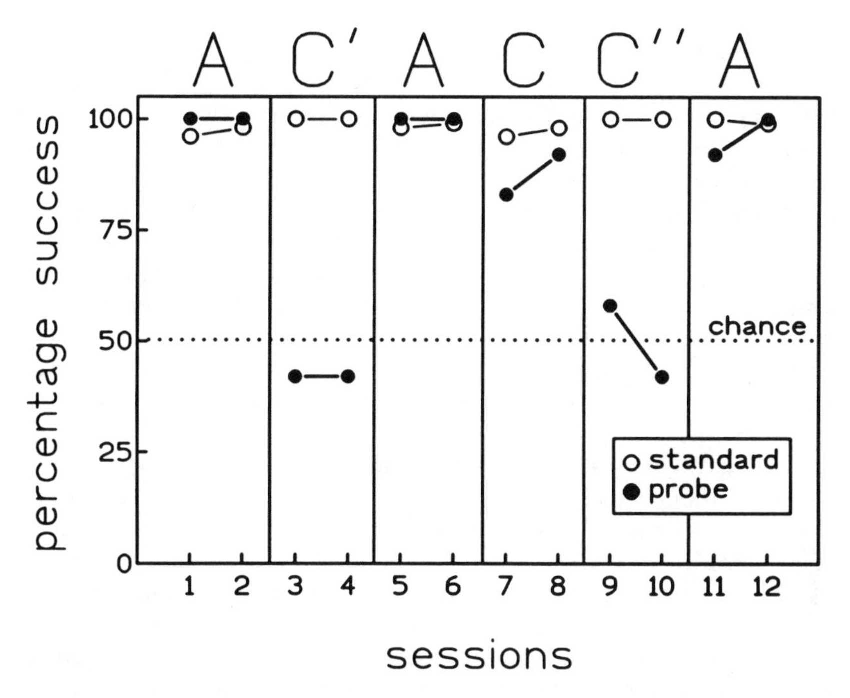

FIG. 8.—Overall results by session, Experiment 3. For details of the treatments, see the text.

= 50%) were conducted for the probe trials in each of the treatment phases (C′, C, C″) and revealed that the subjects did not perform above chance in either the hands-over-eyes or the looking-over-shoulder phases. However, consistent with the results of the first two studies, the subjects performed significantly above chance in the back-versus-front phase ($t[5] = 6.71$, $p = .001$). Finally, a one-way repeated-measures ANOVA confirmed the visual impressions from Figure 8 by detecting an overall effect of phase ($F[5, 25] = 20.0$, $p < .0001$), with Tukey-Kramer post hoc tests revealing that none of the baselines differed from each other or the front-versus-back treatment, whereas all these treatments differed significantly from both the hands-over-eyes and the looking-over-shoulder treatments ($p < .01$ or smaller in all cases). The post hoc tests also revealed that the subjects' chance-level performance during the hands-over-eyes and looking-over-shoulder treatments did not differ.

The two raters who scored the videotapes of the probe trials for hesitations agreed on 99% of the 144 trials that they observed. The average percentage of trials on which the subjects hesitated in each of the six phases ranged from 0% to 27%, but a one-way repeated-measures ANOVA did not yield an overall effect.

These results reveal that the subjects responded to the hands-over-eyes (C′) treatment by gesturing without regard to whether their gesture could be seen by the experimenter. In addition, the subjects also showed a strong preference for the experimenter who faced forward. These two findings suggest that the performance of the subjects in the final test session of Experiment 2 did not reflect a learning curve on the part of the chimpanzees with respect to the hands-over-eyes condition, nor do they suggest that the chimpanzees had "unlearned" their disposition for the experimenter facing forward. Rather, they suggest the viability of the interpretation that four probe trials of mixed types within each session caused the subjects to anticipate poor performance and hence rely on idiosyncratic patterns of responding, such as choosing the right or left response hole by default on most trials.

Perhaps the most striking result of this experiment was that the subjects gestured without regard to whether the experimenters were looking over their shoulder or looking away from them. This finding is consistent with the view that the subjects were not attending to the attentional link between the experimenters' visual access and the gestures they produced. In addition, this result also casts strong doubt on the interpretation that, in the blindfold and bucket condition of Experiment 1, the subjects thought that the experimenter whose eyes were obscured was peeking. In the looking-over-shoulder treatment, we made "peeking" clear and explicit, yet the subjects still gestured without regard to the visual access of the experimenters.

EXPERIMENT 4

At this point, the subjects had shown little evidence that they understood the necessity of the experimenters' visual access to their gesturing. In two of three "naturalistic" treatments (C′, C″), the subjects showed no disposition for gesturing toward the experimenter who could see them. In addition, in both of the conditions involving objects as causes of visual occlusion, the subjects showed no disposition for gesturing toward the experimenter with visual access to what was happening. In the next experiment, we focused on our results from the treatments involving objects and attempted to rule out certain procedural explanations of the subjects' failure to show a preference for the visually unimpeded experimenter. We reasoned that the mentalistic framework could explain the chimpanzees' chance-level performance by arguing that the subjects were startled when they entered the test room and saw blindfolds over the eyes and mouths of the two experimenters or buckets over their heads or on their shoulders. Even though we had given them access to these materials for several days ahead of time, seeing these objects on the experimenters may have altered their emotional state, which in turn could have interfered with their ability to demonstrate an (existing) comprehension of the task. The relatively high percentage of trials in which the subjects displayed visible hesitations in the blindfold and bucket conditions in Experiments 1–3 is consistent with this view, although other interpretations are possible as well.

We decided to test this emotional arousal hypothesis by introducing the blindfolds in a way that did not result in visual obstruction into a treatment (back-versus-front) in which the subjects had repeatedly demonstrated a preference for the visually unimpeded experimenter. Thus, the behaviorist framework predicted that the introduction of the blindfolds into the back-versus-front treatment would have no effect on the subjects' performance; in contrast, the mentalistic framework's appeal to an emotional arousal effect caused by the blindfolds predicted random performance on this mixed treatment.

Method

Subjects

The same six chimpanzees used in the previous studies participated in the present one. They began the experiment the day after the completion of Experiment 3.

Procedure

The general procedure was the same as in Experiments 1 and 3, with 10 trials per session, consisting of eight standard gesturing trials and two probe trials on Trials 5 and 10. The animals were tested using an A-B-C+B-A design. In conditions A and B, the probe trial treatments were identical to the previous experiments (baseline and blindfolds, respectively). We tested the subjects on B first to be sure that the subjects still did not understand the blindfold treatment before we attempted to determine whether the cause of their chance performance was the simple presence of the blindfolds. By mixing treatments (C+B) in the next phase, we combined a treatment for which they previously had a disposition to select the forward-facing experimenter with the (irrelevant) presence of the blindfolds. In this new treatment, one experimenter faced forward, the other faced backward, and both wore a blindfold around the mouth (see Fig. 9). Thus, one of the trainers could see the subject, the other could not, but both were associated with a blindfold. The return to baseline was to ensure that deviations from successful performance were due to the application of the treatments, not to unrelated temporal changes in motivation on the part of

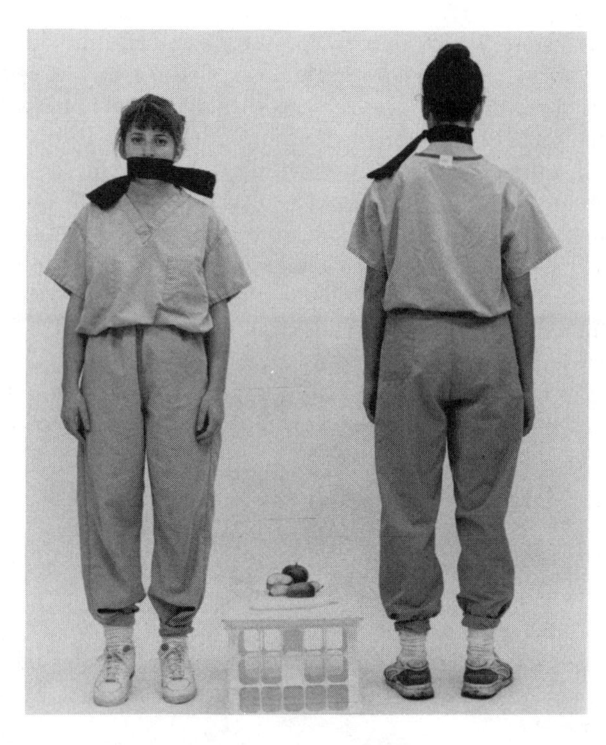

FIG. 9.—Stimulus configuration for the mixed treatment of blindfolds plus back-versus-front (C+B) used in Experiment 4.

the subjects. All randomization and counterbalancing of variables, coding of videotaped probe trials, data summaries, and analyses were conducted as described in Experiment 1.

Results and Discussion

The main results are presented in Figure 10. As in Experiments 1 and 3, the subjects' performance on the standard trials was nearly flawless (over 98% correct). Also as previously, the subjects performed at chance levels in the blindfold condition. However, in contrast to the prediction of the behaviorist framework, and in support of the emotional arousal hypothesis, Figure 10 reveals that the group failed to perform at above-chance levels in the mixed treatment phase (C+B). One-sample *t* tests (two tailed) confirmed the visual impression of Figure 10 by revealing that the subjects' performance did not differ from a hypothetical value of 50% (chance) in either treatment (B, C+B). A one-way repeated-measures ANOVA con-

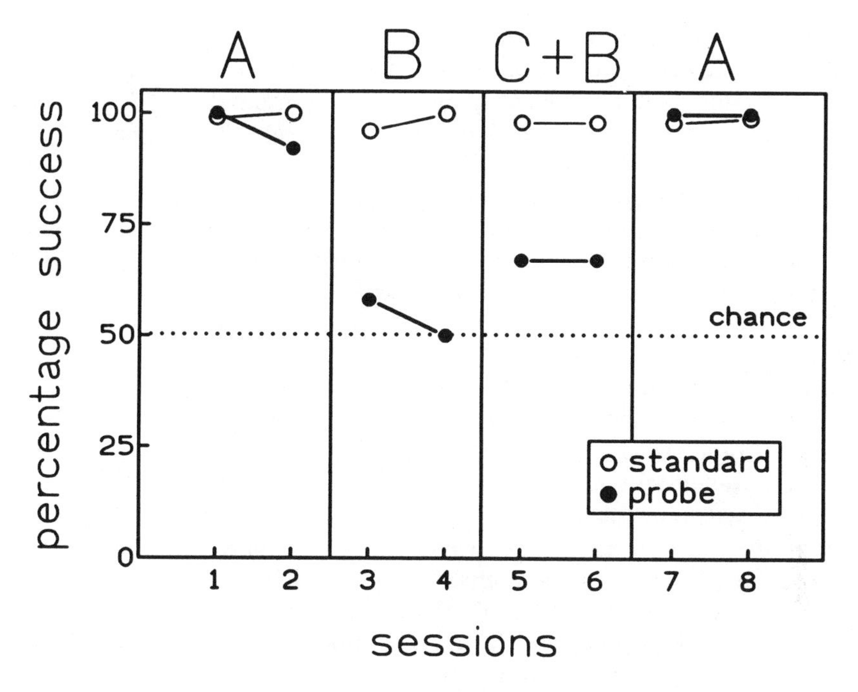

FIG. 10.—Overall results by session, Experiment 4. For details of the treatments, see the text.

firmed that there was an overall effect of phase ($F[3, 15] = 27.69$, $p < .0001$), with each of the two baseline phases differing significantly from each of the two treatment phases (Tukey-Kramer post hoc tests, $p < .001$ in all cases). Post hoc tests also confirmed that the two treatments (B, C + B) did not differ from each other.

The two observers who scored the probe trials for hesitations agreed on 88% of the 95 trials that they rated (one trial was not available owing to a taping error). The average percentage of trials on which the subjects hesitated ranged from 2% to 21% across the four phases of the experiments. A one-way repeated-measures ANOVA revealed an overall effect of phase ($F[3, 15] = 5.18$, $p = .01$). As in Experiment 2, post hoc Tukey-Kramer tests revealed that the effect was due to the fact that the subjects hesitated significantly more often on the blindfold only (B) probe trials than on the probe trials in each of the other three conditions ($p < .05$ in all cases).

An inspection of the pattern of results in Figure 10 illustrates that the subjects still showed no preference for the experimenter who could see them in the blindfold treatment. This was important to establish because, before undergoing this treatment, they had each received a total of eight probe trials involving the blindfolded experimenters and may have learned the correct choice through trial and error. Thus, these results provide a baseline with which we can be certain that the subjects either (*a*) still did not understand the attentional significance of the blindfolds that covered one of the experimenter's eyes or, conversely, (*b*) still were emotionally aroused by the presence of the blindfolds.

To our surprise, in the mixed treatment phase (C+B), the subjects' performances continued to remain at chance levels. This is consistent with the view that the presence of the blindfolds (even without resulting in visual obstruction) interfered with the subjects' ability to perform a task they previously understood. It is worth noting that one subject (Megan) was correct on 100% (four of four) of her C+B probe trials. Because of the nature of the subjects' default tendencies on probe trials they failed to understand (typically choosing all one side), we tentatively regarded Megan's performance as demonstrating that for her the blindfolds did not interfere with her comprehension of the back-versus-front treatment. In no previous treatments had we observed a subject perform 100% correct, except in treatments in which the group's performance did not differ from baseline.

Although the emotional arousal hypothesis was supported, the results of this experiment are in fact consistent with two views. One, of course, is the previously discussed possibility that the subjects understood the previous manipulations involving objects as instruments of visual occlusion but that the presence of those objects interfered with their ability to demonstrate comprehension. This view, however, does not explain the subjects' poor performance on two of the three tasks involving no objects (hands-over-eyes and looking-over-shoulder). The second view is that the results of the mixed treatment phase (C+B) of this experiment can be attributed to the blindfolds having become a conditioned stimulus signaling a probe trial on which the subjects were incorrect half the time. Consistent with this view is the fact that the phases of baseline trials that involved novel objects (blocks of wood) initially contained fairly high levels of hesitations (17%) by the subjects but that hesitations progressively declined across repeated presentations (0%–4%). In contrast, phases of blindfold trials showed an increase from 23% to 56% between Experiments 1 and 2, and hesitations were still five times higher (at 21%) in the current experiment than the treatment phase that contained the next highest amount of hesitations (4%). Either possibility suggested that it was a mistake to include a cue (i.e., the blindfolds) that had previously been associated with failure.

EXPERIMENT 5

In the next experiment, we presented the subjects with experimenters whose vision was again obstructed by an object: a cardboard screen. However, in this case, we thoroughly habituated the subjects to seeing the experimenters with these screens in the context of sessions of standard trials as well as baseline trials before the critical probe trials were administered. We reasoned that associating the screens with successful trials would allow us to choose intelligently between the two possible explanations of the outcome of Experiment 4. First, the behaviorist framework predicted that, if the subjects were habituated to these cardboard screens in the context of correct performance on standard gesturing and baseline trials, then they should not interfere with the back-versus-front-with-screen treatment (because they would not be associated with incorrect performance). The framework also predicted that, when the screens were used as a means of visual occlusion, the subjects should perform at chance levels (because it is assumed that they do not appreciate the mental significance of visual perception). In contrast, if the emotional arousal hypothesis (as applied to the mentalistic framework) were correct, then prior habituation to the objects should result in successful performance on both the back-versus-front-with-screen treatment and the visual occlusion (screen-over-face) treatment.

Method

Subjects

The subjects were the same six chimpanzees that participated in the previous four experiments. Training and testing began 3 days after the completion of the previous experiment.

Materials

Prior to the beginning of this study, several opaque screens were constructed from cardboard. The screens were circular and had a diameter of 30 cm. Nylon cords were fastened in a loop to the screens so that they could be worn around the necks of the trainer and experimenters without interfering with their movements. In addition, in preparation for the experiment, several cardboard boxes were placed in the subjects' home cage, and they were allowed to play freely with them.

Procedure for Pretest Orientation to the Screens

In order to habituate the subjects to the screens, and in an effort to associate the screens with successful performance, we administered three 10-trial sessions of standard trials to each subject with the trainer and experimenters simply wearing the screens around their necks. Thus, on each trial, only one experimenter was in front of the Lexan partition, and the subjects were rewarded by that experimenter after the subject entered the testing unit and gestured correctly. No probe trials were administered. In addition, the trainer also played with the subjects both outside and inside their cages with the screens by alternately holding a screen up in front of his face (and their faces) and taking it away.

Testing Procedure

The general procedure was the same as in Experiments 1, 3, and 4, with 10 trials per session, each session consisting of eight standard gesturing trials and probe trials on Trials 5 and 10. The animals were tested using an A-C_1-A-C_2-B″-C_2-A design. The critical portion of the design was the local C_2-B″-C_2 portion of the testing, which was targeted for a traditional ABA analysis. The probe trials in condition A consisted of the usual baseline but with both experimenters holding the screens up above their left shoulder and offering the block or food with their right hand. The probe trials in C_1 consisted of the usual back-versus-front treatment with the screens hanging around the experimenters' necks as in the training trials. The purpose of this was gradually to ease the subjects into successful performance on the back-versus-front treatment with the addition of the screens. In C_2, the back-versus-front treatment was used with the experimenters holding the screens above their shoulders as in the baseline treatment. In the critical treatment B″ (screen-over-face), both experimenters faced forward, with one holding the screen above her or his shoulder and the other holding it so that it completely obscured his or her face (Fig. 11).

All other details of design, data collection, and randomization were the same as in previous experiments, except that the experimenters now stared at a point on the Lexan partition (designated by a very small mark) about .75 m higher than the previous target. We included this change in an attempt to make their eyes even more salient to the subjects.

Results and Discussion

The subjects performed without a single error across the three pretest sessions of 10 standard gesturing trials in which the cardboard screens were hanging around the experimenters' necks.

The overall results of the main experiment are presented by phase in Figure 12. As previously, the subjects performed over 98% correct on standard trials, providing a within-session control on their motivation to respond correctly. With respect to probe trial performance, the pattern of the results through treatment B″ is consistent with the predictions derived from

FIG. 11.—Stimulus configuration for the screen-over-face (B″) treatment used in Experiments 5–6 (and Experiments 7–9, Chap. IV).

the behaviorist framework. The subjects performed well on the baseline and back-versus-front probe trials (except, inexplicably, on the second session of C_2), and their performance dropped to chance levels in the two sessions of the screen-over-face treatment (B″). However, the group's performance failed to rebound in the second C_2 phase. Separate one-sample t tests (two tailed, hypothetical mean = 50%) for each treatment phase revealed that the subjects' performances were significantly above chance in C_1 ($t[5] = 6.71, p = .001$) and the first C_2 phase ($t[5] = 4, p = .01$). In contrast, the group's performance did not differ from chance in the critical screen-over-face phase (B″) or the back-versus-front phase (C_2) immediately following it. A one-way repeated-measures ANOVA detected a significant overall effect of phase ($F[6, 30] = 6.28, p = .0002$), which Tukey-Kramer post hoc tests indicated was due to the fact that performances in the screen-over-face phase and the second C_2 phase were both significantly lower than performances in each of the baseline phases ($p < .05$ in all six cases). None of the baseline phases differed from each other.

The two observers who scored the probe trials for evidence of hesitations agreed on 95% of the 166 trials that they observed (two of the trials

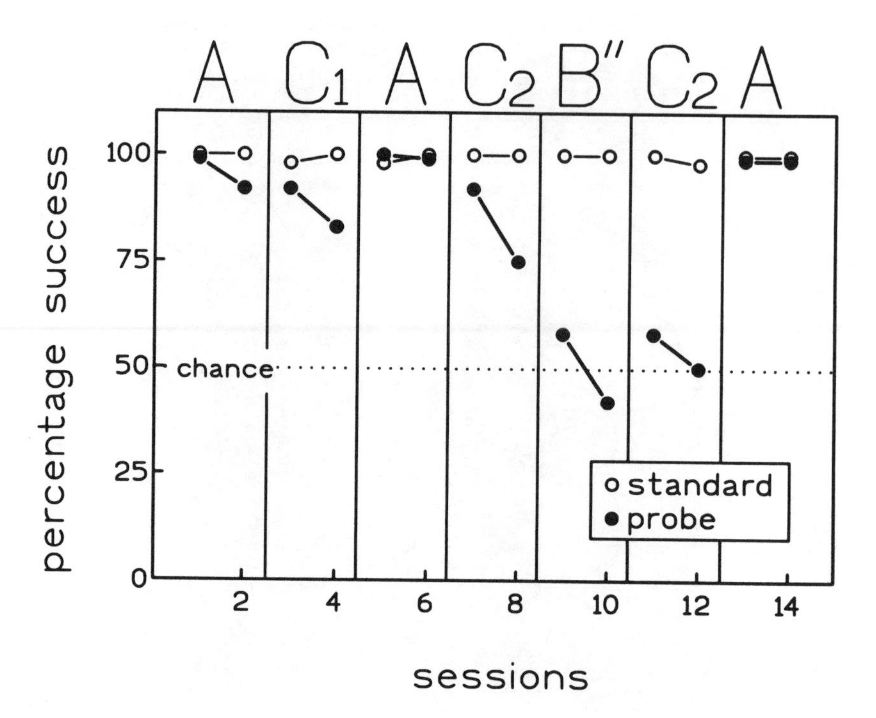

FIG. 12.—Overall results by session of Experiment 5. For details of the treatments, see the text.

were not available for scoring owing to taping errors). The average percentage of trials on which the subjects hesitated ranged from a low of 0% during the second baseline phase to a high of 15% during the back-versus-front phase following the screen-over-face phase. The subjects hesitated the same amount during the first C_2 (back-versus-front) phase as they did during the screen-over-face (B″) phase. A one-way repeated-measures ANOVA indicated no overall significant effect. This is of interest because, unlike in other object treatments in previous experiments, we did not confound the presence of a "novel" object with the introduction of a novel probe treatment itself. Thus, the absence of a hesitation spike in B″ suggests that we had been successful in establishing the neutrality of the cardboard screen. The results of this experiment are difficult to interpret given the subjects' odd performance in the relevant local ABA portion of the experiment (C_2-B″-C_2). The subjects' failure to rebound in the final C_2 phase makes it difficult to attribute the decline in the visual occlusion phase (B″) to the treatment itself. However, the immediate and dramatic rebound in the final baseline phase suggested an explanation for this effect. We reasoned that perhaps the subjects were very sensitive to poor performance in the treatment phases and were updating their expectations about their performance on the basis of cues related to the objects used by the experimenters such as the blindfolds or screens. Thus, their failure in the screen-over-face phase may have led the subjects to anticipate poor performance in the presence of two experimenters holding up screens. In other words, perhaps we had been successful in establishing the neutrality of the screens but simply failed to anticipate the association between objects and failure that might occur within the screen-over-face phase itself. Such an updating of expectations may have caused the subjects to respond using default strategies (i.e., side preferences) in the second back-versus-front phase, thus resulting in chance performance. Our previous experiments had shown that, although they performed at above-chance levels in this treatment, they by no means performed flawlessly. The previous studies also revealed that the subjects virtually always performed correctly on baseline probe trials. We therefore reasoned that, if baseline sessions were interspersed between all treatment phases, we could reset the subjects' expectations about their performances in the presence of two experimenters holding up screens.

EXPERIMENT 6

In the next experiment, we explicitly tested the model outlined in the discussion of Experiment 5 by modifying the previous investigation by interspersing baseline phases between each treatment phase. We reasoned that, if the subjects were not sensitive to the visual deprivation instantiated in the screen-over-face phase, we could reset their expectations before each treatment phase by administering intervening baseline phases and thereby demonstrate the ABA effect originally anticipated by the behaviorist framework at the outset of Experiment 5.

Method

Subjects

The subjects were the same six chimpanzees used in the previous investigation. They began the current study the day after completing Experiment 5.

Orientation Procedure

On the first day of the experiment, each subject received two sessions of 10 trials in which only one experimenter was present on each trial and the experimenter held the screen up above the right shoulder.

Testing Procedure

The general procedures were the same as in Experiments 1 and 3–5. In this experiment, an ABA design was used in which each treatment phase was both preceded and followed by a baseline phase: A-C_2-A-B″-A-C_2-A. Additionally, in contrast to Experiment 5, the experimenters held the cardboard screens above their shoulders on all standard and probe trials. We included the back-versus-front phase with screens held up (C_2) for the same reason as in Experiment 5: as an additional control to ensure that, in the event of poor performance in the visual deprivation phase (B″), we could demonstrate that the subjects had still been motivated and attending on probe trials and were not distracted by the sudden presence of the screens being held up.

Results and Discussion

Orientation

The subjects performed without a single error on the 20 orientation trials they each received in which a single experimenter was present and held up a screen over his or her shoulder.

Testing

The subjects made only a single mistake on the probe trials during the four baseline phases, consistent with their previous high performance on this type of probe trial. In addition, the subjects performed 99% correct on the standard trials averaged across all phases. Given their flawless performance on the baseline probe trials as well as the standard trials, only the critical results of the C_2-B″-C_2 portion of the experiment are considered further.

The main results are depicted in Figure 13. This overall pattern confirms the model discussed in Experiment 5. One-sample two-tailed t tests (hypothetical mean = 50%) confirmed that the subjects performed significantly above chance in the first and second back-versus-front phases ($t[5] = 2.91$, $p = .03$, and $t[5] = 11$, $p = .0001$, respectively) but not in the screen-over-face phase. A one-way repeated-measures ANOVA yielded an overall significant effect across the seven phases of the experiment ($F[6, 30] = 10.42$, $p = .0001$), and post hoc Tukey-Kramer tests revealed that the screen-over-face phase differed significantly from every baseline phase ($p < .05$ or smaller in all six cases), none of which differed from each other. In addition, the screen-over-face phase also differed from the second back-

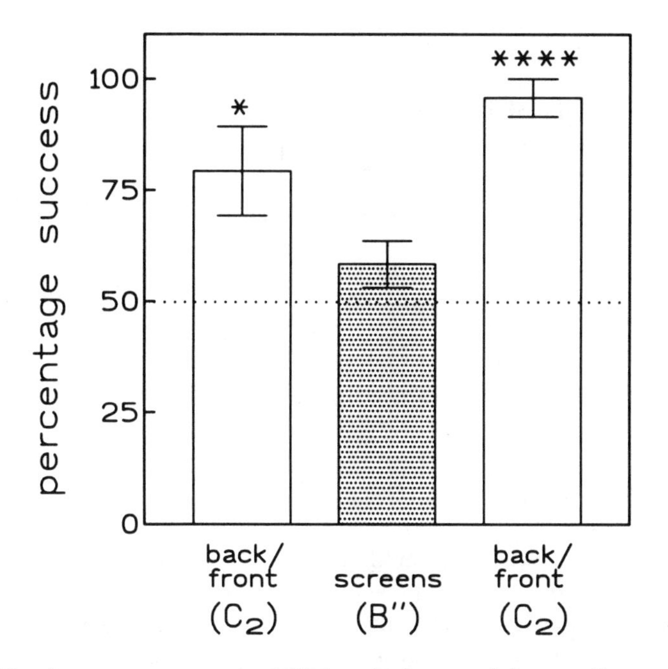

Fig. 13.—Percentage success (± SEM [standard error of the mean]) on probe trials in Experiment 6. Baseline phases preceded and followed each phase of probe trials. The dotted line indicates the level of performance expected by chance responding. * $p < .05$. **** $p < .0001$.

versus-front phase ($p < .001$) and approached a significant difference from the first ($p \approx .06$).

The two observers who separately scored the videotapes of the probe trials agreed as to whether the subjects hesitated before making a choice on 98% of the 168 trials they scored. The average percentage of trials during which the subjects hesitated by phase varied from a low of 0% in the second baseline phase to 12% in the third baseline and second back-versus-front treatment. A one-way repeated-measures ANOVA revealed no overall effect. As in Experiment 5, this absence of a hesitation spike associated with the screen-over-face phase suggests that the low performance in the phase was not an artifact of some sudden change in the subjects' emotional state.

The results of this experiment confirm both the apparent absence of understanding of the visual deprivation manipulation caused by the cardboard screens and the model of the subjects' reactions to failure described in the previous experiment. Apparently, the subjects were extremely sensitive to failure on probe trials for which they had no clear response disposition. However, our results also demonstrate that such failure-induced expectations could easily be reset by interspersing phases of successful probe trials (baseline) before and after the treatments. It is curious that the subjects were so sensitive to failure on the probe trials given that they were typically successful on eight of 10 trials (the standard trials) in every session. However, those trials involved only a single experimenter. Thus, it appears that the subjects parsed the trials within a session into those that involved a single experimenter and those that involved two of them. Thus, success on the standard trials does not appear to have interacted with their expectations concerning probe trials. This suggests that, although our efforts to keep the subjects motivated by using the easy standard trials may have achieved that end, they did not preclude the subjects from interpreting trials in which one experimenter was present in a very different manner from those in which two were present.

SUMMARY AND GENERAL DISCUSSION OF EXPERIMENTS 1–6

The results of all probe trials (except the mixed C + B treatment in Experiment 4) from Experiments 1–6 are summarized by subject in Table 3. The pattern that emerges from this meta-analysis is clear. The subjects displayed an immediate, unlearned disposition in only two types of probe trials: back-versus-front and baseline. In these treatments, the subjects performed at above-chance levels from their first two trials forward, and thus their success cannot be attributed to learning occurring within the context of these experiments. Nonetheless, despite their success in the back-versus-front condition, we believe that the overall results are most consistent with the behaviorist framework outlined at the beginning of this chapter. In all three of the conditions in which the subjects were confronted with visual occlusion caused by objects (blindfolds, buckets, screens), the subjects did not gesture selectively in front of the experimenter who could see them. The same result was obtained in two of the three conditions in which visual occlusion was instantiated by more natural means (hands-over-eyes, looking-over-shoulder). Thus, in only one treatment, back-versus-front, did the subjects show a significant preference for the experimenter who was visually connected to the situation. In addition, the results of the looking-over-shoulder condition from Experiment 3 indicated that this preference may have been due to the frontal stimulus of the experimenters, not the visual contact per se.

Although the results are remarkably consistent across the six experiments, we identified several aspects of the general procedure that might have compromised our test of the mentalistic framework. In particular, three aspects of how we presented the two experimenters to the chimpanzees may have interfered with their ability to demonstrate a sensitivity to the mental significance of the visual obstructions. First, the experimenters'

TABLE 3

SUMMARY OF BASELINE AND TREATMENT PERFORMANCES BY INDIVIDUAL SUBJECTS, EXPERIMENTS 1–6

		OBJECT CONDITIONS			NATURAL CONDITIONS		
SUBJECT	BASELINE	Blindfold	Bucket	Screen-over-Face	Back-versus-Front	Hands-over-Eyes	Looking-over-Shoulder
Kara	63/64	6/12	3/4	6/8	31/36	5/12	2/4
Jadine	64/64	5/12	2/4	4/8	32/36	7/12	3/4
Brandy	62/64	6/12	3/4	2/8	27/36	6/12	1/4
Megan	62/64	6/12	2/4	5/8	34/36	6/12	2/4
Mindy	62/64	6/12	1/4	4/8	29/36	8/12	2/4
Apollo	64/64	6/12	3/4	5/8	28/36	6/12	2/4
M	.98	.49	.58	.54	.84	.53	.50

NOTE.—Data summarized across Experiments 1–6.

faces were considerably higher than the chimpanzees' eye level. Although the subjects' extremely high performance on the baseline and back-versus-front trials demonstrates that they looked up at some point after the trainer let them into the testing room, we cannot say with certainty that they carefully focused on the experimenters' faces before making their choices. Second, the experimenters stood 1.5 m in front of the subjects. It is possible that, at this distance, they failed to attend as carefully as they might have had the experimenters been closer. A final concern centered around the gaze of the experimenters. We chose not to have the experimenters stare into the eyes of the chimpanzees, for reasons discussed in Experiment 1. However, having them stare at a neutral position on the Lexan partition midway between the response holes was perhaps inferior to having them stare at the holes directly in front of them. It is possible that the subjects initially scanned the experimenters' faces and determined that neither one was looking at them or at the location where they would make their response and hence discounted the visual gaze information in making their choices.

IV. UNDERSTANDING WHO CAN SEE YOU: FURTHER INVESTIGATIONS

In this chapter, we report a series of studies that we designed and conducted to obtain additional information before provisionally choosing between the mentalistic and the behaviorist explanations of young chimpanzees' understanding of visual perception. As noted at the end of Chapter II, several aspects of the design of our original studies may have masked an underlying mentalistic appreciation of seeing on the part of our young chimpanzees. In the experiments reported in this chapter, we attempted to assess the robustness of the chimpanzees' apparent failure to comprehend the significance of the fact that only one of the experimenters was subjectively connected to them via visual perception. To this end, we conducted a series of experiments in which we attempted to make the visual gaze of the experimenters more salient to the subjects. In most of the studies, we positioned the experimenters in a manner that would make their visual gaze as easy for the subjects to detect as possible. In Experiments 7–9, we used one of the treatments using objects as means of visual occlusion, and, in Experiments 10–12, we tested more naturalistic instances of visual occlusion and inattention. Finally, in Experiments 13 and 14, we implemented several treatments to test a learning model of their performance.

EXPERIMENT 7

The purpose of the first study was to address the possibility outlined at the end of Chapter III—that the subjects' apparent lack of comprehension on the critical probe trials (other than back-versus-front) may have been due to procedural aspects of how the experimenters were presented to the subjects. Ideally, naive subjects would have been used, but this was impossible given the practical limitations of obtaining and training new subjects. Nonetheless, we adopted an approach that we believed could minimize the problems associated with repeated exposure to the trials. In this experiment, we altered several variables simultaneously in order to maximize the chances that, if the subjects did appreciate the mental significance of seeing, they would demonstrate it. Thus, in this study, we made no explicit attempt to test for the relative importance of each of the variables altered. Rather, we sought to alter the configuration of the experimenters in order to make their differential visual access as obvious as possible.

Method

Subjects

The subjects were the same six chimpanzees that participated in the experiments described in Chapter III. At the beginning of this study, the subjects ranged in age from 5-1 to 5-9. The current experiment began 2 days after the completion of Experiment 6.

Procedure

The overall design of the procedure was identical to that of Experiment 6 in Chapter III: an ABA design (back-versus-front, screen-over-face, back-versus-front) with baseline phases both preceding and following each treatment phase (i.e., A-C_2-A-B″-A-C_2-A). However, three aspects of the manner in which the experimenters were presented to the subjects were altered: their height/posture, their proximity to the front of the partition, and the orientation of their gaze. Instead of standing 1.5 m away from the testing unit and staring at a neutral point on the panel, the experimenters sat on upside-down milk crates 0.75 m in front of the panel and stared at the response hole directly in front of them. The purpose of these alterations was (*a*) to make eye gaze more *salient* by moving the experimenters closer and positioning their faces at a height more comparable to that of the subjects' faces when they were responding and (*b*) to make the gaze direction more *relevant* by having each experimenter stare at the response hole directly in front of him or her. This setup is shown in Figure 14 in the context of a back-versus-front trial (without screens).

One effect of having the experimenters sit down was to leave us with a decision on the back-versus-front trials concerning whether to have the main trunk of both experimenters' bodies equally distant from the testing unit, but with the feet of the experimenter facing forward slightly closer. The alternative was to stagger the distance of the crates so that the feet of the person facing forward were in line with the main trunk of the experimenter facing backward. In the end, we decided to have the trunks of their bodies equally distant. It is important to note, however, that at this point

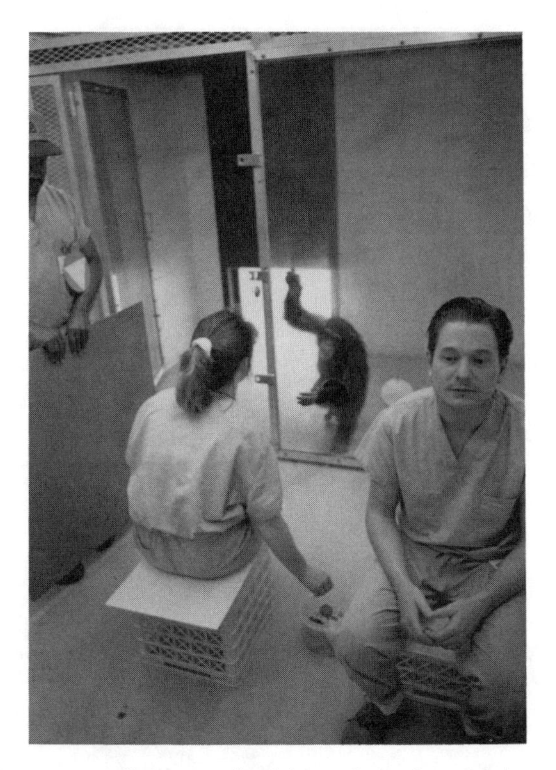

FIG. 14.—New posture, distance, and height configuration of the experimenters used in Experiments 7–13. Note that in the back-versus-front condition (depicted here) the feet of the experimenter facing forward are closer to the test panel than those of the experimenter facing away; however, the main trunk and head of both experimenters are equally close to the panel (approximately 0.75 m).

we were no longer particularly interested in the chimpanzees' comprehension of the back-versus-front treatment (by this point they had each received 36 trials of this treatment and performed above chance from Trial 1 forward). Whatever conclusions we could draw from this treatment were already secure using a configuration in which the experimenters stood equally distant from the subjects. From this point forward, these probe trials were used as a basis for controlling for motivation across sessions. We used the same procedures for randomization and counterbalancing and coding of videotapes, which were described in the previous chapter.

Results and Discussion

Consistent with previous findings, the subjects made very few errors on the standard gesturing trials across all phases (over 99% correct). In

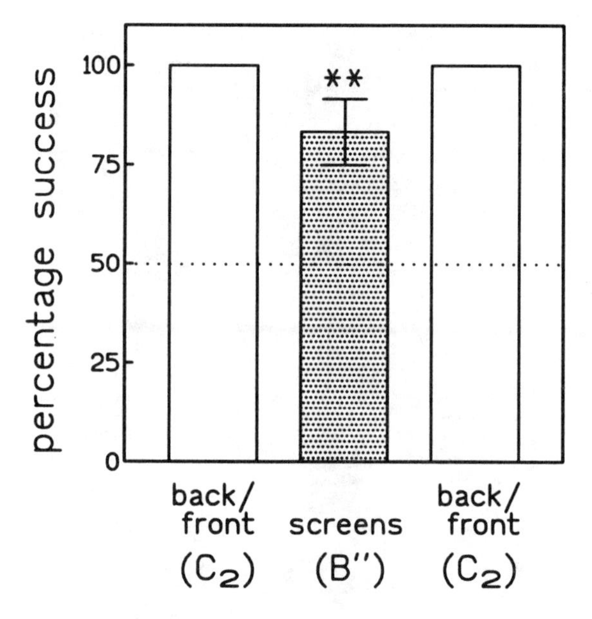

FIG. 15.—Percentage success (± SEM) on probe trials in Experiment 7. Baseline phases preceded and followed each phase of probe trials. The dotted line indicates the level of performance expected by chance responding. ** $p < .01$.

addition, they made no errors on the baseline probe trials. These results are therefore not considered further.

The crucial results of the ABA component (C_2-B''-C_2) of the study are presented in Figure 15. As before, the subjects performed quite well during both the first and the second back-versus-front phases. In fact, no animal made any errors on these probe trials during this experiment—a slightly better than usual performance when compared to performance levels in previous experiments. However, in striking contrast to our previous findings using the cardboard screen as a means of visual occlusion, in this experiment the chimpanzees performed significantly above chance in the screen-over-face phase (one-sample t test, $t[5] = 4.00$, $p = .0103$, two tailed, hypothetical mean = 50%). A one-way repeated-measures ANOVA failed to yield a significant difference among the three phases.

The two observers who separately scored all the probe trials for evidence of hesitations agreed on 90% of the 166 trials they viewed (two trials were not available owing to taping errors). The subjects' hesitations across phases ranged from a low of 2% to a high of 33% of the trials. A one-way repeated-measures ANOVA indicated an overall effect ($F[6, 30] = 5.685$, $p = .0005$), and Tukey-Kramer post hoc tests revealed that this effect was due to the fact that the subjects hesitated significantly more (33% of the probe trials) in the screen-over-face phase than in all other phases, except the first back-versus-front phase, during which they hesitated on 21% of the trials ($p < .05$ or smaller in all cases). No other phases differed significantly from each other.

The significant spike in hesitations associated exclusively with the screen-over-face treatment was surprising, especially when coupled with the absence of a comparable spike in Experiments 5 and 6. Recall that our hypothesis had been that such spikes might be indicative of an emotional arousal or alarm at seeing a novel stimulus configuration. Indeed, in all previous cases, the hesitation spikes were associated with poor performance on the part of the subjects. However, in this case, the hesitation spike was associated with overall correct performance, suggesting a dissociability between hesitation spikes and poor performance. This suggested that the subjects' hesitations might derive from two sources: changes in their cognitive or perceptual states and/or changes in their emotional states.

Although the results of this study suggest that the mentalistic framework might have some heuristic value, it is important to note that by the end of this experiment the subjects had each received a total of 12 probe trials using screens as a means of visual occlusion. Thus, the subjects might have learned a rule unrelated to the mental significance of the visual deprivation: for example, *Avoid the person whose face is covered by the screen.* Although our subjects had not shown evidence of similar learning in other treatments that they initially did not understand in the same (or greater) number of trials (i.e., blindfolds and hands-over-eyes), this did not exclude the possibility that they had simply learned a discrimination in this specific treatment across the repeated trials. This possibility seemed especially likely in light of the fact that we took great pains to accustom the animals to the presence of the screens in the test setting.

The purpose of the current study was not to determine with certainty whether above-chance performance was the direct result of some or all of the alterations we employed. However, a clear method of testing whether the improved performance was caused by our manipulations (e.g., having the experimenters sit closer) suggested itself: conduct an experiment in which half the probe trials involved experimenters sitting close and half involved them standing as in Chapter III. If the subjects performed poorly when responding to experimenters standing at a distance, but succeeded when they were sitting closer, doubt would be cast on the behaviorist account. However, before attempting to tease this issue apart, we felt that it would be necessary to replicate the finding after slight (presumably irrelevant) alterations in the procedure. This seemed especially warranted given that in two previous experiments we had failed to find evidence that the subjects performed successfully in the screen-over-face treatment.

EXPERIMENT 8

Method

Subjects

The subjects were the same six chimpanzees used in all previous studies. They began the current study the day after completing Experiment 7.

Procedure

As in Experiment 7, the experimenters sat on crates 0.5 m away from the testing unit and stared at the response hole in front of them. The rest of the procedure was altered from the prior experiment in several ways. The subjects were tested in sessions consisting of six trials each. Trials 1–2 and 4–5 were designated as spacer trials, but unlike previous experiments they consisted of baseline trials (A) where the subjects choose between two experimenters, one offering food and one offering a block of wood. Trials 3 and 6 were designated as probe trials. Thus, unlike previous experiments, there were no standard gesturing trials.

The subjects were tested using an ABA design: B''-C_2-B''. Sessions 1–2 (Phase 1) contained probe trials of the screen-over-face treatment, Sessions 3–4 (Phase 2) contained probe trials of the back-versus-front treatment, and Sessions 5–6 (Phase 3) again contained probe trials of the screen-over-face treatment. Thus, unlike in previous experiments, there were no sessions containing baseline probe trials between each of the treatment phases. Instead, baseline trials surrounded the probe trials within each session. We utilized the same procedures for randomization and counterbalancing as those used in previous studies. The basic procedure for scoring the videotapes was the same as in previous studies except that all trials within each session were scored for hesitations, not just the probe trials.

Results and Discussion

During the baseline trials surrounding the treatment probe trials, the subjects responded as usual with near-perfect performance (98%, 100%, and 98% correct in Phases 1, 2, and 3, respectively). As usual, these results demonstrate that the subjects were perfectly motivated to participate throughout the entire experiment and are therefore not considered further.

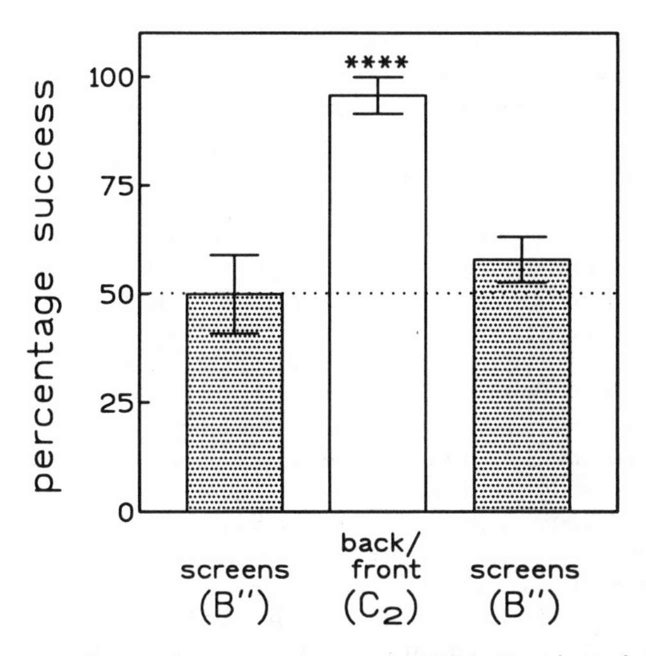

FIG. 16.—Percentage success (± SEM) on probe trials in Experiment 8. The dotted line indicates the level of performance expected by chance responding. **** $p < .0001$.

The main results are presented in Figure 16. As in the previous study, the subjects responded with near-perfect accuracy on the probe trials during the two sessions of the back-versus-front treatment. In striking contrast, they responded at chance levels on the probe trials contained in the screen-over-face phases that occurred before and after the back-versus-front phase. One-sample t tests (two tailed, hypothetical mean = 50%) confirmed the visual impression of Figure 16: the subjects' performance did not differ from chance in either of the screen-over-face phases but did so strongly in the back-versus-front phase ($t[5] = 11$, $p = .0001$). A one-way repeated-measures ANOVA yielded an overall effect of phase ($F[2, 10] = 22.39$, $p = .0002$). Tukey-Kramer tests revealed that the first and second screen-over-face phases each differed significantly from the back-versus-front phase ($p < .001$ and $p < .01$, respectively). The two screen-over-face phases did not differ from each other.

The two observers who scored the tapes for hesitations observed 216 trials, 72 of which were probe trials and 144 of which were baseline trials. The observers agreed with each other on 97% of the baseline trials and 92% of the probe trials. The percentage of probe trials accompanied by hesitations ranged from 4% in the back-versus-front phase to 10% in the second phase of screen-over-face. A one-way repeated-measures ANOVA indicated no overall effect of treatment phase. Interestingly, this absence of a hesitation spike was associated with overall low performance in the screen-over-face phases, just as in Experiments 5 and 6, exactly the opposite pattern of Experiment 7. This suggests that there was no dramatic change in the emotional-cognitive state of the subjects during the screen-over-face probe trials.

The general results of this study are consistent with the view that the subjects were not attending to the significance of the experimenter whose face was obscured by a screen. This is especially evident from several trials in which the subjects not only gestured toward the person whose face was obscured and looked at him or her while doing so but also, after waiting briefly, extended their arm even further through the response hole, as if reexecuting (or emphasizing) the gesture. Such behavior suggests a means-ends understanding of the situation from the chimpanzees' perspective. The subjects appeared to understand the causal relation between their gesture through the response hole and the subsequent behavior of the experimenter reaching down, getting a food reward, and handing it to them. Yet their inattention to the absence of visual perception suggests that they did not understand that a subjective connection (through the eyes) from the experimenter to them was a prerequisite for the experimenter to respond to the gesture.

It is difficult to know why the subjects performed so poorly on the screen-over-face probe trials in this experiment as compared to Experiment 7. At any rate, it was possible to rule out a simple motivational account; as in the previous studies, the subjects' performance was excellent on all trials within the screen-over-face sessions except the critical probe trials themselves. Thus, accounting for the difference between these results and those in Experiment 7 by recourse to some local motivational factors is inadequate. On the other hand, there were some very minor procedural differences between Experiments 7 and 8. The standard gesturing trials were replaced by baseline (A) trials, and we eliminated the sessions between the back-versus-front and the screen-over-face conditions in which there were baseline probe trials. However, the subjects showed no carryover effects from their poor performance in the screen-over-face sessions to the subsequent back-versus-front session as they had in Experiment 5. Thus, the ABA design was quite effective in demonstrating the absence of either a temporal effect in motivation or a carryover effect from the critical treatment phase (screen-over-face). The only other difference was the order in which the treatments were presented; C_2-B″-C_2 in Experiment 7 versus B″-C_2-B″ in Experiment 8. Again, however, the fact that the ABA design was effective offers little room for accounting for the differences on the basis of this procedural alteration.

EXPERIMENT 9

Before attempting to explain the apparently contradictory results obtained in Experiments 7 and 8, we decided that it was necessary to attempt to replicate the finding of an absence of a sensitivity to the screen-over-face treatment (the results from Experiment 8) using the C_2-B''-C_2 design of Experiment 7 but otherwise employing the exact same methods as Experiment 8. This would further help establish whether the findings of Experiment 7 were anomalous or whether the subjects were perhaps demonstrating some form of learning. Before the study began, each subject had received 20 trials using the screens treatment.

Method

Subjects

The subjects included the same six chimpanzees used in all previous studies. An additional subject, Candy (age 5-0), was also tested using the same procedures. Candy had been a member of the subjects' original nursery group from which they had been transferred. She had been introduced into the subjects' social group 7 months before the current experiment (see the "Subjects" section of Experiment 1 in Chap. III above). Since that point, she had participated in daily sessions of gesturing training, receiving the same training procedures as were administered to the other subjects (described in Experiment 1). She had also participated in several unrelated studies. Once Candy had reached criterion on standard trials (see Table 2 above), she was tested on the baseline procedure as well as the screen-over-face and back-versus-front treatments using the same procedures described for the other subjects. When she began the current study, she had received a total of 24 baseline trials, 16 back-versus-front trials, and four screen-over-face trials. Her responses were essentially the same as those of the other subjects. (A summary of the exact designs and results of the preliminary studies can be obtained from the authors.) Before beginning this study, we decided not to include her results on this experiment in the group results but rather to use her performance as a control for determining the potential learning effects that may have been operating with the other subjects. The current study began the day after the completion of Experiment 8.

Procedure

The subjects were tested using the same within-session design used in Experiment 8. The sole difference from Experiment 8 was the across-

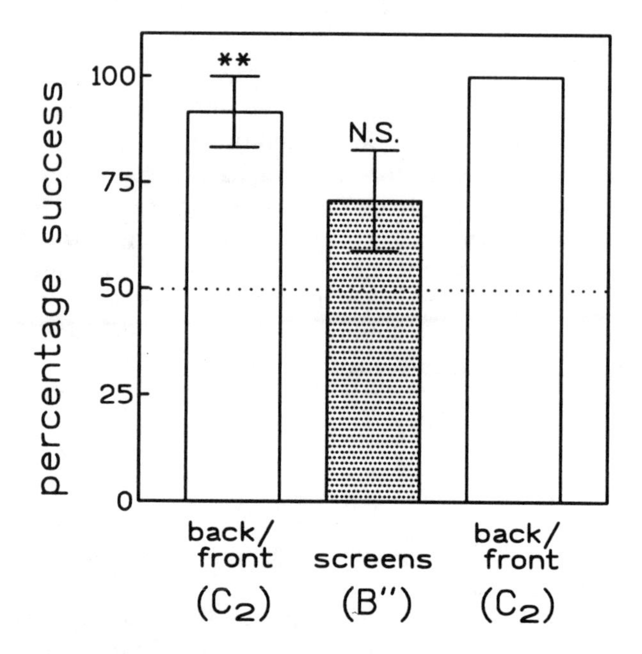

FIG. 17.—Percentage success (± SEM) on probe trials in Experiment 9. The dotted line indicates the level of performance expected by chance responding. ** $p < .01$. N.S. = not significant.

session arrangement of treatments. The subjects were again tested using an ABA design, but the order of treatments was inverted to C_2-B''-C_2. Thus, Sessions 1–2 (Phase 1) contained probe trials of the back-versus-front treatment, Sessions 3–4 (Phase 2) contained probe trials of the screen-over-face treatment, and Sessions 5–6 (Phase 3) again contained probe trials of the back-versus-front treatment. The randomization and counterbalancing procedures were the same as in previous studies, and the videotapes were scored for hesitations on all trials as in previous experiments.

Results and Discussion

As usual, the six main subjects performed extremely well on the baseline trials (99% correct as a group across all six sessions combined). The control subject, Candy, averaged 92% correct on the baseline trials across all sessions of the experiment.

The critical results are depicted in Figure 17 and show the main group's mean performance on the treatment probe trials by condition. The group was correct on an average of 71% of the trials in the screen-over-face phase, but one-sample t tests (two tailed, hypothetical mean = 50%, df = 5) revealed that this performance was not significantly above chance. In contrast, the group again performed almost flawlessly on the probe trials during the two phases of the back-versus-front treatment. A one-way repeated-measures ANOVA approached an overall effect of treatment phase ($F[2, 10] = 3.10, p = .089$). In general, these results are consistent with the findings from Experiments 5, 6, and 8, but not Experiment 7, in indicating the absence of a clear disposition for the subjects selectively to gesture toward the experimenter whose face was fully visible. The results of the control subject, Candy, are consistent with the predictions of the behaviorist framework: 100% correct on the first back-versus-front phase, 50% on the screen-over-face phase, and 100% on the second back-versus-front phase.

The two observers who separately scored the videotapes of the probe trials agreed as to whether the subjects hesitated before making a choice on 96% of the 72 probe trials and on over 99% of the baseline trials. The average percentage of probe trials during which the subjects hesitated ranged from a low of 2% in the second back-versus-front phase to a high of only 6% in both the first back-versus-front and the screen-over-face phase. A one-way repeated-measures ANOVA revealed no overall effect. As in the previous experiments, we interpret the absence of an increase in hesitations during the screens phase as evidence that the subjects' poor performance was not due to an emotional arousal caused by seeing that one of the experimenter's faces was missing.

Taken as a whole, the results of Experiments 7–9 provide little data to support the view that chimpanzees understand the intentional property of visual perception, despite our alterations in the procedure. However, Experiment 7 stands out in that, unlike the other four studies, the subjects did perform significantly above chance in the screen-over-face phase. We feel that the best interpretation of these results is that the subjects were gradually learning to avoid the person whose face was obscured by the screen but had not yet done so with any consistent reliability. Their variable performance (including 71% correct in the current study) suggests that the subjects were in the process of learning such a rule (see Experiments 13 and 14 below). Finally, our inability to demonstrate a stable discrimination coincident with the alteration of the procedures left us with little reason to test the idea that it was some aspect of the procedure (distance, height, gaze direction), and not learning per se, that had elevated the subjects' performance in this specific treatment in Experiment 7.

EXPERIMENT 10

Despite our inability to demonstrate a stable, successful performance in response to the alteration of the distance, height, and gaze of the experimenters in one of the object conditions (screen-over-face), the possibility was still open that these alterations might have made a difference in the context of one of the naturalistic treatments. Thus, we selected the looking-over-shoulder (C″) treatment for further analysis using the new procedures outlined in this chapter. We chose looking-over-shoulder for two reasons. First, of the remaining two naturalistic treatments, we believed that looking-over-shoulder was the clearest example of one experimenter being visually connected to the situation and the other not. Equally important, the six original subjects had each received only four trials using this treatment, and those trials had been administered 3 months prior to the beginning of this experiment. The subjects performed at chance (50% correct) or lower, except Jadine, who performed at 75% (three of four) correct (see Table 3 above). In addition, Candy had received no probe trials of this nature, thus providing an additional control who was completely naive with respect to this treatment.

Method

Subjects

The subjects were the seven chimpanzees used in the previous investigations. Before the experiment began, we decided to include the additional subject, Candy, in the overall experiment and average her data along with those of the others, although we were also interested in her data alone. Thus, except where noted, all subsequent group means and data analyses are based on all seven subjects combined. This experiment began the day after Experiment 9 was completed.

Procedure

The structure of the individual testing sessions was the same as in the previous experiment: four baseline spacing trials and two probe trials. The design of the treatment phases was as follows: C-C″-C-C″-C-C″. As usual, each phase consisted of two sessions. The C sessions contained back-versus-front probe trials. On the probe trials in the C″ sessions, the experimenters both faced away from the front of the testing unit, and one experimenter turned her or his head to look back at the hole in the Lexan partition. The other experimenter oriented his or her shoulders in the same manner but continued to look away so that only the back of his or her head was visible from inside the testing unit (see Fig. 7 above). All randomization, counterbalancing, and videotape scoring procedures continued as in the previous experiments.

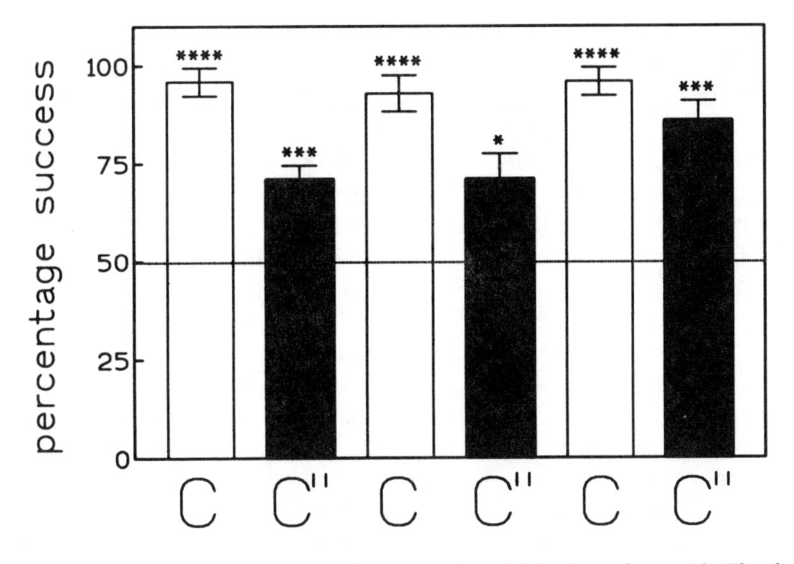

Fig. 18.—Percentage success (± SEM) on probe trials in Experiment 10. The dotted line indicates the level of performance expected by chance responding. * $p < .05$. *** $p < .001$. **** $p < .0001$.

Results and Discussion

As usual, the subjects performed excellently on the baseline trials, averaging 97% correct across all sessions. These results are therefore not considered further.

The main results of the experiment are presented in Figure 18. The subjects responded with near-perfect accuracy during each of the three back-versus-front phases. In contrast, the subjects averaged 71%, 71%, and 86% correct during the three looking-over-shoulder phases. A one-way repeated-measures ANOVA revealed a highly significant overall effect ($F[5, 30] = 8.24$, $p < .0001$). Tukey-Kramer post hoc tests revealed that none of the back-versus-front phases or the final looking-over-shoulder phases differed from each other. However, the subjects' performance in the first and second looking-over-shoulder phases was significantly lower than in all back-versus-front phases ($p < .05$ or smaller in all cases). None of the looking-over-shoulder phases differed from each other. Thus, in general, the subjects performed significantly worse on the looking-over-shoulder probe trials than on the back-versus-front trials. However, one-sample t tests (two tailed, hypothetical mean = 50%) indicated that the subjects nonetheless performed significantly above chance in all three looking-over-shoulder phases ($df = 6$, all t's ≥ 2.828 and ≤ 13.000, all p's $< .02$ or smaller). Thus, although the subjects had a more difficult time with the looking-over-shoulder probe trials, they did display a statistically significant preference for gesturing in front of the person who could see them. This effect is also true for Candy, the subject who had never experienced the treatment before: 75%, 75%, and 75% correct for each of the three phases of C″.

As usual, the two observers who scored the videotapes for hesitations displayed high levels of interobserver reliability, agreeing with each other on 96% of the 504 baseline and probe trials that they coded. More specifically, they agreed as to whether the subjects hesitated before making a choice on 96% of the 168 probe trials that they scored separately. Because each subject received the same number of probe trials in each of the two conditions in an alternating fashion, for purposes of analysis each subject's 12 probe trials in each condition were averaged. Next, a paired t test was conducted comparing average percentage of trials with hesitations in the two conditions. As a group, the subjects hesitated on 11.8% of the looking-over-shoulder probe trials as compared to only 1.8% of the back-versus-front trials ($t[6] = 2.884, p = .027$). This effect indicates a strong association of the introduction of the treatment C″, which the subjects had experienced on just four trials several months earlier, and a slowdown in their responses. Note that these results mirror those found in Experiment 7 (unlike the other experiments) in that the subjects hesitated most frequently on novel probe trials on which they performed fairly well.

The results of this experiment are not consistent with those reported in Experiment 3, in which the subjects responded at chance on the four trials that each subject received of the looking-over-shoulder treatment. In this experiment, the subjects responded at above-chance levels on the first block of four trials. Even if the one-sample t test analysis is restricted to the first two trials, the subjects still performed significantly above chance ($t[6] = 2.83, p = .03$). Indeed, even better evidence that the response was not learned in the context of this experiment was that all seven subjects responded correctly on Trial 1, a result significantly greater than that expected by chance responding (binomial test, $p = .0078$). Because the correct choices were counterbalanced both within sessions (between subjects) and between sessions (within subjects), this impressive Trial 1 performance could not be due to some unaccounted for preference for one side over the other: three of the subjects' correct choices were on the right side, and the other four subjects' correct choices were on the left side.

We entertained two possible explanations of the results of this study. First, the results could be interpreted as evidence consistent with the predictions generated by the mentalistic framework. That is, from Trial 1 forward, they gestured in front of the experimenter who could see them at a level that exceeded that expected by chance responding. However, six of the subjects had already received four trials of this type prior to this experiment (Experiment 3 above), although in that case the experimenters were standing up and were further away from the subjects. Nonetheless, it is possible that the subjects learned the correct contingency extremely rapidly and transferred it to this experiment.

The individual performance of two subjects may cast doubt on this explanation. First, Jadine never responded below 75% correct in any phase using the looking-over-shoulder treatment: 75%, 75%, 100%, and 100% across the four blocks of four probe trials that she received across the two experiments. In addition, Candy had received no previous trials of this type before this experiment, and she performed at 75% correct across the three phases of this experiment). Although this does not rule out the learning theory interpretation, it does suggest that perhaps this experiment (unlike Experiments 7–9) revealed an effect of manipulating the distance, height, and/or gaze direction of the experimenters. In other words, perhaps the subjects performed better in this experiment than in Experiment 3, not because they had learned a rule about the situation, but because the manner of presenting the experimenters elicited a disposition that had previously been masked for procedural reasons.

EXPERIMENT 11

In this experiment, we sought to test the hypothesis that the subjects' poorer performance on the looking-over-shoulder treatment in Experiment 3 as compared with Experiment 10 was due to the height, proximity, and/or gaze direction of the experimenters between whom the subjects chose. We thus used a design in which on half the probe trials of looking-over-shoulder the experimenters were positioned as in Experiment 3 and on the other half they were positioned as in Experiment 10. In addition, we also manipulated the eye gaze of the experimenters so that on half the trials both experimenters fixed their gaze on the same target location on the Lexan partition as in Experiment 3; on the other half they looked just above their respective holes where the subjects responded (as in Experiments 7–10). If the subjects' Trial 1 above-chance responding in Experiment 10 was the result of a learning effect independent of our alterations of the manner in which the experimenters were presented, then their performance should be above chance in both the far and the near conditions. On the other hand, if the subjects' above-chance performance in Experiment 10 was due to the unmasking of a disposition to gesture toward the experimenter who could see them by making the faces more salient, then their performance should be high in the near condition but at chance levels in the far condition.

Method

Subjects

The subjects were the same seven subjects used in the previous investigation. They began the current experiment 4 days after completing Experiment 10.

Procedure

The subjects were tested using the design detailed in Table 4. Each subject was tested across nine phases consisting of two sessions each. The design for all subjects was C-C″-C-C″-C-C″-C-C″-C. In order to explore the effect that distance and height had on the subjects' performance, the subjects each received eight of their 16 looking-over-shoulder probe trials using the original procedures from Experiment 3 with the experimenters standing up and further away from the testing unit. On the other eight probe trials, they sat on the crates as in Experiment 10. To explore the effects of gaze direction, the subjects received four of their eight "near" looking-over-shoulder probe trials looking at the target above the middle hole as in the earlier experiments (Position 1) and four looking at the response hole as in the current series of studies (Position 2). The "far" looking-over-shoulder probe trials were divided in the same fashion. In order to control for possible effects of testing order, we counterbalanced the order in which the near and far C″ probe trials were administered by dividing the subjects into two groups (Group 1 = Kara, Jadine, Brandy; Group 2 = Megan, Mindy, Apollo, Candy), with Group 1 receiving the near C″ treatment first and alternating thereafter and Group 2 alternating in the opposite order. Table 4 provides additional details of the assignment of gaze direction to the various treatments.

For this experiment, additional randomization procedures were utilized that altered some of the counterbalancing procedures imposed earlier. The purpose of the changes was to further challenge local "win-stay" and "lose-shift" strategies that the subjects were apparently using across trials that they did not understand. For these schedules, we no longer required that the correct choice appear equally often on both sides within each session. However, the sides were correct equally often across sessions. In addition, we altered the constraint that the same side could not be correct on more than three consecutive trials; this was changed so that a given side could be correct from zero to six times within each session. However, the two sessions within each phase were still counterbalanced so that each location was correct equally often.

TABLE 4

EXPERIMENTAL DESIGN AND RESULTS FOR EXPERIMENT 11

SUBJECT	SESSIONS AND CONDITIONS								
	1–2	3–4	5–6	7–8	9–10	11–12	13–14	15–16	17–18
Group 1									
Kara	C (near) 100%	C″ (near,1,2) 100%	C (near) 100%	C″ (far,2,1) 100%	C (near) 100%	C″ (near,1,2) 100%	C (near) 100%	C″ (far,2,1) 50%	C (near) 100%
Jadine	C (near) 100%	C″ (near,2,1) 100%	C (near) 100%	C″ (far,1,2) 100%	C (near) 100%	C″ (near,2,1) 100%	C (near) 100%	C″ (far,1,2) 100%	C (near) 100%
Brandy	C (near) 100%	C″ (near,1,2) 100%	C (near) 100%	C″ (far,2,1) 100%	C (near) 100%	C″ (near,1,2) 75%	C (near) 100%	C″ (far,2,1) 75%	C (near) 100%
Group 2									
Megan	C (near) 100%	C″ (far, 2,1) 50%	C (near) 100%	C″ (near,1,2) 100%	C (near) 100%	C″ (far,2,1) 75%	C (near) 100%	C″ (near,1,2) 100%	C (near) 100%
Mindy	C (near) 100%	C″ (far,1,2) 50%	C (near) 100%	C″ (near,2,1) 100%	C (near) 100%	C″ (far,1,2) 100%	C (near) 100%	C″ (near,2,1) 75%	C (near) 100%
Apollo	C (near) 100%	C″ (far,2,1) 75%	C (near) 100%	C″ (near,1,2) 75%	C (near) 100%	C″ (far,2,1) 50%	C (near) 100%	C″ (near,1,2) 100%	C (near) 100%
Candy	C (near) 100%	C″ (far,1,2) 100%	C (near) 100%	C″ (near,2,1) 50%	C (near) 100%	C″ (far,1,2) 50%	C (near) 100%	C″ (near,2,1) 100%	C (near) 100%

NOTE.—C = back-versus-front; C″ = looking-over-shoulder; near = both experimenters sitting down as in Experiments 7–10; far = both experimenters standing up as in Experiments 1–6; 1 = experimenters fix gaze at target as in Experiments 5–6; 2 = both experimenters fix gaze on response hole in front of them as in Experiments 7–10; first number indicates eye gaze position for first session within phase; second number indicates eye gaze position for second position within each phase.

Results and Discussion

As usual, the subjects responded at near-perfect levels on the baseline trials in all sessions (99% correct overall). These results are therefore not considered further.

The results of the treatment probe trials by subject are presented in Table 4. Overall, the subjects averaged 89% correct on the C″ probe trials in which the experimenters sat near the Lexan partition and 79% correct when they stood further away. One-sample *t* tests (two tailed, hypothetical mean = 50%) revealed that, in both the near and the far C″ conditions, the subjects performed above what would be expected by chance ($t[6]$ = 7.778 and 5.435, p = .0002 and .0016, respectively). A 2 (distance: near, far) × 2 (gaze direction: Position 1, Position 2) factorial ANOVA was conducted and revealed no effect of distance or gaze nor an interaction between the two. Thus, there appeared to be little support for the hypothesis that distance or eye gaze, by themselves, had an effect on the subjects' tendency to gesture in front of the person who was looking over his or her shoulder. Indeed, the absence of an effect of eye gaze in the direction expected (see the introduction to Experiment 7 above) was particularly striking: the subjects averaged 87% correct when the subjects looked at Position 1 (the position used in the early studies) but only 80% when they looked at Position 2 (the position used in later studies), and the same trend held true if the means of the near and far trials are examined separately (near: 93% for Position 1 vs. 86% for Position 2; far: 82% for Position 1 vs. 75% for Position 2). This nonsignificant trend was in the opposite direction expected by the mentalistic framework's post hoc explanation of the differences between the subjects' performances on looking-over-shoulder between Experiments 3 and 10. In contrast, the trends for distance are more suggestive that distance may have had an effect in the direction predicted that was too small to detect: the subjects averaged 89% correct for near trials and only 79% for far trials. Again, the same trend held if Position 1 versus Position 2 eye gaze trials are examined separately (Position 1: 93% for far vs. 82% for near; Position 2: 86% for far vs. 75% for near).

The observers who scored the videotapes for hesitations on the probe trials agreed with each other on 96% of the 250 probe trials that were available for analysis (two trials were missing owing to taping errors). The percentage of trials in which the subjects hesitated ranged from 13.4% in the far condition of looking-over-shoulder to a low of 0% in the fourth back-versus-front phase. A one-way repeated-measures ANOVA indicated no overall effect. The absence of a significant difference in hesitations across treatments is important because it is consistent with the data from their primary responses indicating that the subjects did not differentiate among the two critical conditions, the far and near versions of looking-over-shoulder. Thus, this is additional evidence that the variables that we manipulated (height, distance, eye gaze) were not the source of the subjects' difficulty with this treatment in Experiment 3 or of their success in Experiment 10.

The results of this experiment are not consistent with the mentalistic framework's account of the subjects' differential performance in the looking-over-shoulder treatment in Experiments 3 and 10. If poor performance in Experiment 3 had been due to the distance, height, or eye gaze direction (or some interaction among these variables), then there ought to have been a clear near versus far or Position 1 versus Position 2 effect. Instead, the results suggested to us a variant of the learning explanation for the differential results obtained in Experiments 3 and 10 in the looking-over-shoulder condition. We reasoned that, instead of assessing each treatment separately, perhaps the subjects were combining all probe trials involving two experimenters into a larger pool and were extracting a rule that could satisfy most treatments.

Perhaps the most salient rule that could cover all the treatments (except blindfolds) would be, *Gesture in front of the person whose face is visible.* Thus, instead of beginning Experiment 10 with only four trials of relevance to the looking-over-shoulder treatment, each subject may have begun having experienced from 156 to 224 probe trials (depending on whether the back-versus-front trials are excluded) in which such a rule would have been successful. Of course, an even more specific rule might have been extracted as well, such as *Gesture in front of the person whose eyes are visible;* this rule would have led to successful performance on the blindfold trials as well. To be fair, however, the mentalistic framework could account for these results in a post hoc fashion by arguing either (1) that the subjects now understood the significance of visual perception or (2) that they understood it all along but demonstrated it only in the most obvious, natural situation (back-versus-front) and, with a little more difficulty, in the second most obvious natural situation (looking-over-shoulder).

However, the relatively naive subject, Candy, represents a potential problem for the learning account interpretation offered above. She performed quite well in Experiment 10 despite having received far fewer total (as well as far fewer types of) probe trials than the other subjects. On the other hand, it is possible that learning was more rapid in the near conditions than in the far conditions, and all Candy's trials had been in the near condition.

EXPERIMENT 12

In the next experiment, we sought to evaluate the alternative explanations discussed above for why the chimpanzees had shown a clear preference for gesturing in front of the person who could see them during the looking-over-shoulder probe trials of Experiments 10 and 11. We created a new naturalistic treatment in which both experimenters sat on the crates and faced forward. One of the experimenters stared straight ahead as usual, but the other looked up above the testing unit, as if distracted. We interpreted the behaviorist rule(s), *Gesture in front of the person whose face/eyes is/are visible,* to predict that the subjects should respond at chance when confronted with this treatment. After all, were the subjects merely using the presence of a face or eyes to determine their responses, then both experimenters should have equal valence because both of their eyes and faces were visible. In contrast, we interpreted the post hoc mentalistic account of the results of Experiments 10 and 11 to predict that the subjects should gesture toward the experimenter who was paying attention. We also predicted in advance that in either case the subjects might respond to the gaze direction of the "distracted" experimenter by turning and looking up to follow his or her gaze (Scaife & Bruner, 1975). Furthermore, if they followed the gaze of the distracted experimenter without selectively gesturing toward the experimenter who was paying attention, this would demonstrate in a rather dramatic way how responding to visual gaze or following line of regard might be independent of an understanding of the subjective dimension of visual perception that we were investigating.

Method

Subjects

The subjects were the same seven chimpanzees used in the previous experiments. They began the current study 2 days after completing Experiment 11.

Phase 1 Procedure

The first phase of the experiment consisted of four sessions, with each session including the four usual spacer baseline (A) trials (Trials 1–2 and 4–5) and two probe trials (Trials 3 and 6). In addition to the usual back-versus-front probe trials, a new naturalistic treatment (C''': attending-versus-distracted) was created to test the hypotheses outlined above. In this new treatment, both experimenters sat on the crates and faced forward with their eyes open, but one of the experimenters raised his or her head and stared at a predetermined location on the ceiling above the testing unit (see Fig. 19). The location was chosen so that, if the subjects noticed and attempted to follow the experimenter's gaze as they entered the testing unit, they would be forced to turn and orient their heads in a manner that would

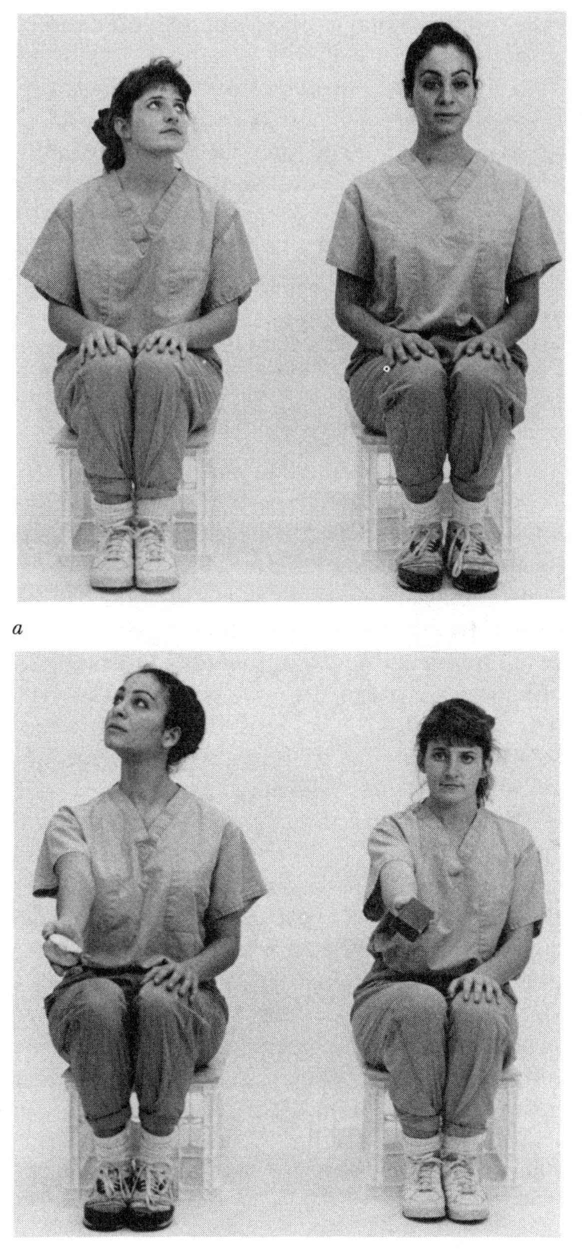

Fig. 19.—The stimulus configuration for experimenters for Experiment 12. *a*, Attending-versus-distracted (C‴). *b*, Baseline-plus-attending-versus-distracted (A′).

be easy to detect on the videotapes. The other experimenter stared straight ahead at the response hole as usual.

In order to mitigate any possible effects of testing order, the subjects were randomly assigned to two groups. Group 1 consisted of Kara, Megan, Apollo, and Candy; Group 2 consisted of Jadine, Brandy, and Mindy. Group 1 received two sessions containing probe trials of attending-versus-distracted followed by two sessions containing probe trials of back-versus-front. Group 2 was administered the treatments in the opposite order. Thus, across four test sessions, all the subjects received four probe trials of attending-versus-distracted and four probe trials of back-versus-front in an order that was counterbalanced between groups.

Phase 2 Procedure

The second phase was planned in the event that the subjects both (*a*) turned to follow the distracted experimenter's gaze and (*b*) performed randomly on the C''' probe trials. Such a finding could implicate the subjects' own distraction in following the distracted experimenter's gaze as the source of their random performance when choosing between the two. In other words, it could be argued that they chose randomly because they were distracted after attending to the experimenter's unusual gaze. In order to test this interpretation, we planned four sessions that each consisted of six baseline (A) trials. In the first two sessions, the probe trials were exactly the same as the spacer trials (baseline trials in which one experimenter held out a block of wood and the other held out a food reward). The purpose of these first two sessions was to rehabituate the subjects to seeing both experimenters staring straight ahead as they typically did. In Sessions 3 and 4, however, the baseline probe trials differed in one respect: the experimenter who held out the food looked up above the testing unit in the same manner as the distracted experimenter had done in Phase 1 (except that the location was on the opposite side of the ceiling in order to minimize habituation from Phase 1). This probe trial type was designated as baseline-plus-attending-versus-distracted (A') (see Fig. 19). The predictions of the two frameworks were clear: if the behaviorist view were correct, the subjects should perform well on the A' trials; if the mentalistic framework were correct and random performance on C''' were obtained, then the subjects should also perform randomly on A'.

Data Analysis

As usual, all trials were coded from videotape for hesitations by two observers. Two additional observers were instructed to view the 335 trials in the experiment that were available (one trial was missing because of a taping error) and to judge whether the subjects glanced up into the corner in which the distracted experimenter looked on the probe trials (this was the left side for Phase 1 and the right side for Phase 2). Thus, the observers coded such glances for all trials, not just the probe trials containing a distracted experimenter, in order to obtain a measure of the ambient level of such glances. The data sheets were constructed so that the raters could note whether the subjects glanced before and/or after gesturing toward one of the experimenters. Thus, this meant that each observer made 670 separate judgments, two per trial: (1) a judgment concerning the subjects' behavior before they gestured and (2) a separate judgment about their behavior after they gestured. After determining that there was an acceptable level of interobserver reliability, the two observers' scores were averaged. From these measures, it was possible to determine whether the subjects selectively glanced up at the ceiling on the distracted-versus-attending probe trials as opposed to other trials. Also, we could determine whether they glanced before or after they gestured toward one of the experimenters.

In addition to this coding, an additional observer, who was naive as to the purpose of the experiment, was used to code whether the subjects looked up to the ceiling on all trials across the experiment. We told this observer that the experimenters had placed a large plastic toy grasshopper on the ceiling above the testing unit in order to determine how attentive the subjects were being during the trials. Because the videotapes did not reveal the ceiling above the testing unit, we were able to claim that the grasshopper was out of sight in the left-hand corner of the ceiling in Phase 1 and the right-hand corner in Phase 2. The naive observer was shown on a paused videotape of the setting approximately where the grasshopper was allegedly located in Phase 1 and Phase 2. We instructed the observer to review each trial and record whether she believed that the chimpanzees had noticed the grasshopper on the ceiling, on the basis of whether the animals turned their heads and looked up to the appropriate location.

Results

As usual, the subjects performed virtually flawlessly on the baseline trials across the two phases of the experiment. Only a single error was made across the eight sessions of testing. As before, these results demonstrate that the subjects were adequately motivated and attending and are therefore not considered further.

Following the Experimenter's Gaze

In order to determine whether the subjects selectively responded to the distracted experimenter's gaze, the frequency of the subjects' glances into

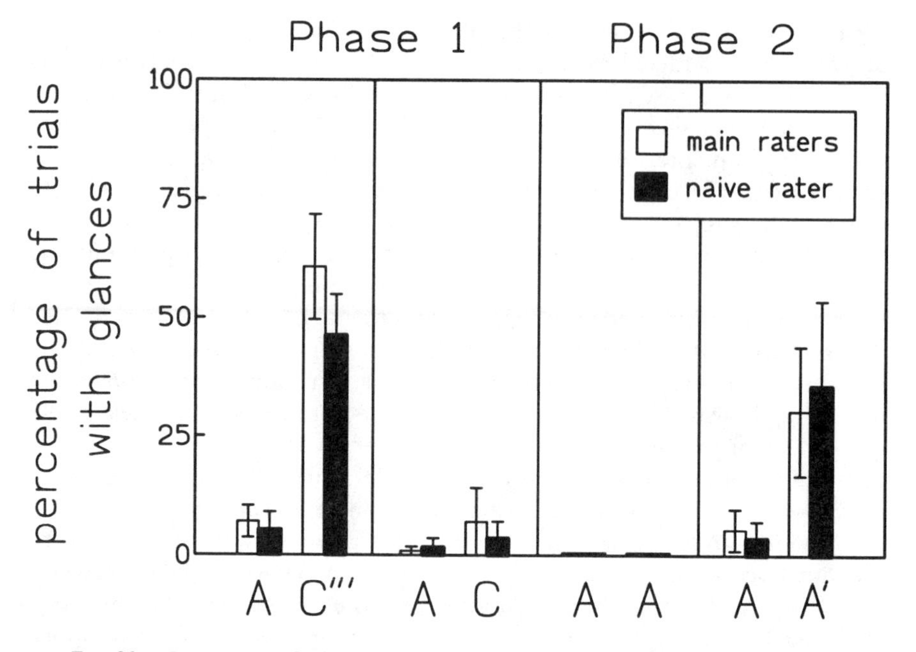

FIG. 20.—Percentage of trials in Experiment 12 in which the chimpanzees responded to the distracted experimenter's gaze by glancing upward to track the distracted experimenter's gaze (± SEM) as scored by main and naive observers. For details of the comparisons, see the text.

the target corner were compared across all trial types. First, the two main observers' scores were analyzed across all trials using the percentage agreement calculation described in previous experiments. Across all judgments (N = 670) that they were required to make, the observers agreed with each other 97% of the time as to whether the subjects glanced up into the corner. Second, the analysis was restricted to the probe trials alone, and the observers still displayed high levels of interobserver reliability, agreeing with each other on 95% of these judgments (N = 222). The two main observers' scores were therefore collapsed to produce an average percentage of trials for each subject in each trial type during which they glanced up into the target corner. A group mean was then obtained by averaging each of the seven animals' scores.

The open bars in Figure 20 present the average percentage of trials during which the group glanced into the target corner of the ceiling across the two phases of the experiment as judged by the two main observers. For purposes of clarity, the glances that occurred during baseline (A) trials that surrounded the probe trials within each session type are displayed separately, adjacent to the glances that occurred during the probe trials themselves. This allows for a direct comparison between the subjects' reactions to the distracted experimenter's gaze and any spontaneously occurring (ambient) levels of glances to that same location. Figure 20 clearly shows that the subjects were extremely sensitive to the distracted experimenter's gaze. In Phase 1, the subjects glanced up in the direction that the experimenter was looking on 61% of all attending-versus-distracted probe trials, which is almost nine times more frequently than in any other trial type. In addition, although it is not obvious from Figure 20 (because the A and C‴ are not shown in temporal relation to each other; see above), a trial-by-trial analysis of the actual order in which the trials were delivered indicated that virtually all the glances that did occur on A trials happened *after* the first C‴ trials. This suggests a possible carryover effect from the probe trials.

The filled bars in Figure 20 display the results of the naive observer who was not told the purpose of the experiment. Her results clearly mirror those obtained by the observers who knew the true purpose of the study. For purposes of statistical analyses, all three observers' scores were averaged, and a one-way, repeated-measures ANOVA was used to compare the percentage of trials during which the subjects glanced up to the target location on all eight trial types/blocks displayed in Figure 20. The results indicated an overall effect of trial type/block ($F[7, 42] = 9.59$, $p < .0001$). Tukey-Kramer post hoc tests indicated that this overall effect was the result of the subjects' glancing to the target location on a significantly higher percentage of trials in the C‴ condition than in all other trial type/blocks, except (as expected) A′ (all p's $< .05$). Likewise, the subjects glanced into the target location significantly more often on A′ trials than in all other trial type/blocks (all p's $< .05$), except for C‴ (as noted above) and C, and the latter comparison approached a statistical difference.

Given that the subjects showed a robust sensitivity to follow the distracted experimenter's gaze, a further analysis of the data was conducted to determine whether the subjects tended to respond to the distracted experimenter's gaze before or after they made their main response of gesturing toward one of the two experimenters. If the subjects visually tracked the distracted experimenter's gaze *before* gesturing toward one of the experimenters, their subsequent performance may have been compromised by forcing them to attend to two variables (the distracted experimenter's gaze and the two options confronting them) at the same time. The data (from all three observers averaged) reveal that the subjects glanced approximately equally often both before and after gesturing toward one of the experimenters: the subjects glanced before gesturing toward an experimenter on 42% of the C‴ trials and/or glanced after gesturing on 36% of the trials; the subjects glanced before gesturing toward an experimenter on 20% of the A′ trials and/or glanced after gesturing on 13% of the trials. Thus, this result leaves open the possibility that the subjects' main performance could

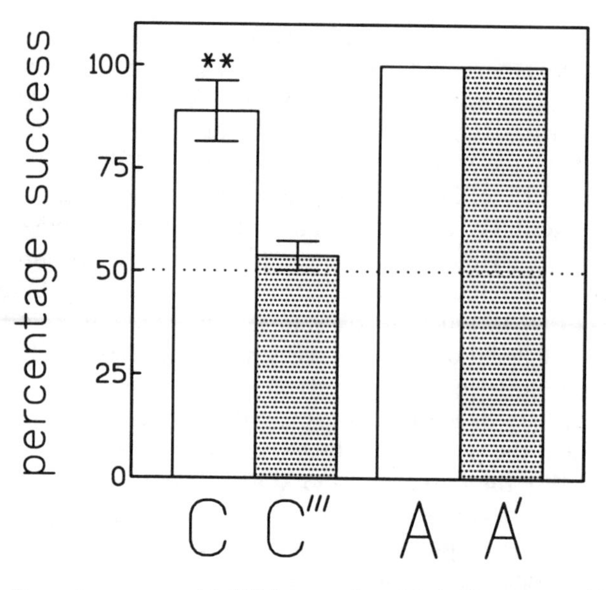

FIG. 21.—Percentage success (± SEM) on probe trials in Experiment 12. The dotted line indicates the level of performance expected by chance responding. ** $p < .01$.

have been affected by their tracking the distracted experimenter's gaze, making the primary responses in Phase 2 critical for interpreting the main effect on the C''' trials.

Responses on Probe Trials

The main results of the probe trials of both phases of the experiment are depicted in Figure 21. The pattern of the subjects' responses clearly matches the predictions generated by the behaviorist framework, not those generated by the mentalistic framework. In Phase 1, the subjects performed excellently on back-versus-front probe trials as usual, responding to the experimenter facing forward at a level well above chance (one-sample t test, two tailed, hypothetical mean = 50%, $t[6] = 5.284$, $p < .002$). In stark contrast, their performance in the attending-versus-distracted session did not differ from chance (one-sample t test, two tailed, hypothetical mean = 50%). A one-way repeated-measures ANOVA indicated that there was an overall effect of probe trial type ($F[3, 18] = 30.400$, $p < .0001$). Tukey-Kramer post hoc tests confirmed that the subjects performed significantly lower on the attending-versus-distracted (C''') trials than all other trial types ($p < .001$ in all cases), none of which differed from each other.

Given that the subjects were glancing up before making their choices on almost half these critical probe trials (see above), the results of Phase 2 are critical for interpreting their random performance. The results of the A′ probe trials in Phase 2 indicate quite clearly that the subjects' tracking of the distracted experimenter's gaze was not the cause of their random performance on the C‴ probe trials; statistically, the subjects glanced up into the target location as much on the A′ probe trials in Phase 2 as they did on the C‴ trials in Phase 1, yet they gestured correctly to the experimenter holding the food on every single A′ probe trial (see Figs. 20 and 21 above).

Additional evidence that the subjects' tracking of the distracted experimenter's gaze was not the cause of their random responses was obtained from a trial-by-trial analysis of the subjects' C‴ probe trials. On the basis of the data from the two main observers, there were 14 cases in which the subjects tracked the experimenter's gaze before responding. In six of those cases, the subjects followed their glance with an incorrect response, and, in eight cases, they followed with a correct response. Likewise, there were 14 cases in which the subjects did not track the experimenter's gaze before responding. In eight of those cases, the subjects were correct, and, in six cases, they were incorrect. Thus, trials on which the subjects did follow the experimenter's gaze and trials on which they did not were equally associated with correct and incorrect responses. This result also demonstrates that the subjects did not perform better when they noticed the distracted experimenter's gaze before they gestured, a finding that is extremely difficult to reconcile with the mentalistic framework.

Hesitations on Probe Trials

The two observers agreed on 94% of all probe trials that they observed as to whether the subjects hesitated before making their gesture. Although a one-way repeated-measures ANOVA failed to indicate an overall effect of probe trial type, the subjects did display their usual pattern of hesitating most on probe trials that were new and/or on which they responded at chance (C‴ = 23%, C = 10%, A = 2%, A′ = 9%).

Discussion

Two critical findings emerged from this experiment. First, the subjects responded randomly when both experimenters' eyes and face were visible, even though one of them was not visually attending to the situation at hand. This effect was present even on those trials for which we have definitive evidence that the subjects noticed the experimenter's distracted gaze. This result exactly matches the prediction generated by the behaviorist framework. The behaviorist interpretation of the subjects' excellent performance in Experiments 10 and 11 on looking-over-shoulder probe trials was that they had learned a rule about gesturing toward the experimenter whose eyes and/or face was visible. Were such a rule all that the subjects understood about the situation, then they would be left with no clear disposition in a case where both experimenters' eyes and faces were present, regardless of any difference in the subjective focus of the two experimenters. These are exactly the results that we obtained.

The second central finding of this experiment is that young chimpanzees will track the visual gaze of another organism (in this case, a member of another species). The general phenomenon of following the gaze of an adult has been demonstrated in human infants by the end of the first year of life (e.g., Butterworth & Cochran, 1980; Butterworth & Jarrett, 1991; Scaife & Bruner, 1975). Although we did not demonstrate that our subjects would respond to the distracted experimenter's eye gaze independent of her or his head orientation, nonetheless, the chimpanzees responded to the general configuration as if they were tracking line of sight. This was an extremely robust phenomenon, with the subjects tracking the distracted experimenter's gaze on over half of all distracted-versus-attending (C‴) probe trials in the first phase of the experiment. However, they did tend to show some habituation by Phase 2, in which they tracked the distracted experimenter's gaze on about one-third of the A′ trials. More recent research in our laboratory has replicated and extended this discovery of gaze following in chimpanzees. For instance, we have shown that 6-year-old chimpanzees will track the gaze of humans in response to both head and eye orientation as well as eye orientation alone (Povinelli & Eddy, in press).

Taken together, these results imply that, even as young chimpanzees automatically track the visual gaze of another organism, they simultaneously appear oblivious as to the attentional implications of that gaze. We interpret this to reflect the development/deployment of an evolutionarily useful behavior, with no attending comprehension of the mental significance of the behavior. This proposal could be quite similar to Butterworth and Jarrett's (1991) notion of an early emerging "ecological" mechanism serving joint visual attention in 6–12-month-old human infants. However, this similarity would have to be reconciled with the fact that 6–12-month-old infants will not track the line of sight of another into space outside their visual field, whereas our chimpanzees clearly did.

We can think of at least two evolutionary contexts in which the tracking of the gaze of another organism might be important. First, for highly social animals, the visual gaze of others may provide important clues about impending social interactions. For example, Chance's (1967) argument concerning the importance of monitoring others in primate societies sets the appropriate stage for an advantage to reading eye gaze to determine who is looking at whom. Of course, rules about head (and possibly just eye) orientation need not include any mentalistic appreciation of seeing.

The second situation in which gaze tracking might be selected for is in the context of predation. Most organisms face predation pressure, and social organisms can obtain some protection from predators by attending to stimuli not available to asocial organisms. Indeed, a major theory for the evolution of sociality is that group living evolved as a response to predation pressure (Alexander, 1974). One proposal is that group living allows for early detection of predators and hence a more rapid escape (van Schaik, van Noordwijk, Warsono, & Sutriono, 1983). One proximate warning signal of such detection could be the sudden shift in the visual monitoring system of another nearby conspecific; such a shift or unusual orientation could serve as a cue for the presence of potential predators. Thus, those organisms that respond to such orientations in others might detect potential predators sooner than those that do not. Given the ubiquity of both of the situations described above, we would not be surprised to discover that the phenomenon is widespread among primates and other highly social species.

Although we comment on this issue further in Chapter VI, it is worth noting here that, if gaze following (including gaze following into space outside one's own field of vision) can occur in chimpanzees or other animals without an understanding of the intentional significance of that gaze, then this raises questions concerning the use of gaze following in human infants as evidence of an understanding of the "aboutness" of visual perception (e.g., Baron-Cohen, 1994).

EXPERIMENT 13

We reasoned that, if the behaviorist account of Experiments 10–12 were correct, then our young chimpanzees ought to perform quite well in a number of conditions in which they had previously performed randomly. In particular, we reasoned that, if the subjects had learned to use the presence of faces or eyes as stimulus cues, then treatments such as screen-over-face, buckets, blindfolds, and hands-over-eyes should no longer necessarily present a problem for them. However, the exact stimulus features available to the subjects differ across conditions, so, depending on the nature of the rule that they had learned, we were able to predict different results. Table 5 presents a summary of the treatments and the predictions of how the subjects should perform if various stimulus features were both necessary and sufficient to elicit a gesture by the subjects. As can be seen, depending on the exact nature of the cues, different predictions are made for the same treatments. Table 5 also includes a new naturalistic treatment, eyes-open-

TABLE 5

PREDICTIONS FOR TREATMENTS USED IN EXPERIMENT 13 BASED ON POTENTIAL LEARNED STIMULUS RULES

CONDITION	POTENTIAL STIMULUS RULES LEARNED BY CHIMPANZEES WITH ATTENDANT PREDICTIONS FOR ABOVE-CHANCE PERFORMANCE		
	Eyes plus Face	Eyes Alone	Face Alone
Natural conditions:			
Eyes-open-versus-closed (C'''')	Yes	Yes	No
Attending-versus-distracted (C''')	No	No	No
Looking-over-shoulder (C'')	Yes	Yes	Yes
Hands-over-eyes (C')	Yes	Yes	Yes?
Object conditions:			
Blindfolds (B)	Yes	Yes	No
Buckets (B')	Yes	Yes	Yes
Screen-over-face (B'')	Yes	Yes	Yes

versus-closed, which we employed as an additional test of the *Gesture toward the person whose eyes are visible* rule. In this case, both experimenters' faces were completely visible, but on one of the experimenter's face the eyes were present, and in the other they were absent. We conducted the following experiment to determine which of the stimulus features seemed to be controlling the subjects' behavior.

Method

Subjects

The subjects were the same seven chimpanzees used in the previous experiments, and they began the current study 2 days after completing Experiment 12.

Procedure

The subjects were tested on seven treatments: blindfolds, buckets, screen-over-face, hands-over-eyes, looking-over-shoulder, distracted-versus-attending, and the new naturalistic treatment, eyes-open-versus-closed (C''''; see Fig. 22). The general structure of the test sessions was the same as in the previous experiment. Each session consisted of six trials, with Trials 1–2 and 4–5 serving as spacer trials, and with Trials 3 and 6 serving as probe trial slots. However, unlike Experiments 8–12, the spacer trials were not baseline (A) trials but were instead back-versus-front trials. Each subject received two sessions containing probe trials of a given treatment, followed

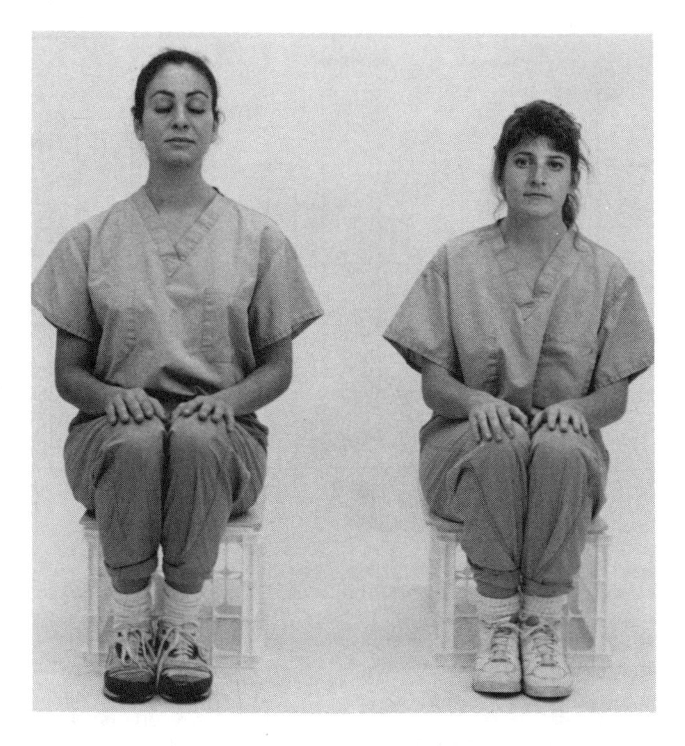

FIG. 22.—Stimulus configuration of experimenters for eyes-open-versus-closed (C'''') treatment used in Experiment 13.

by another two sessions containing probe trials of a different treatment, and so on, until they had received two sessions containing each of the seven treatments. Thus, each animal was tested across 14 sessions and received four probe trials of each of the seven treatments. Each of the seven subjects received a novel testing order for the seven treatments so that across subjects each treatment occurred once in each of the seven possible ordinal positions. All other counterbalancing and randomization procedures followed those described in previous experiments.

Results and Discussion

The subjects performed excellently on the back-versus-front spacer trials that surrounded the probe trials, with their performance ranging from 96% to 100% across the pairs of sessions. Given that these results establish that the subjects were attending and motivated throughout the experiment, they are not considered further.

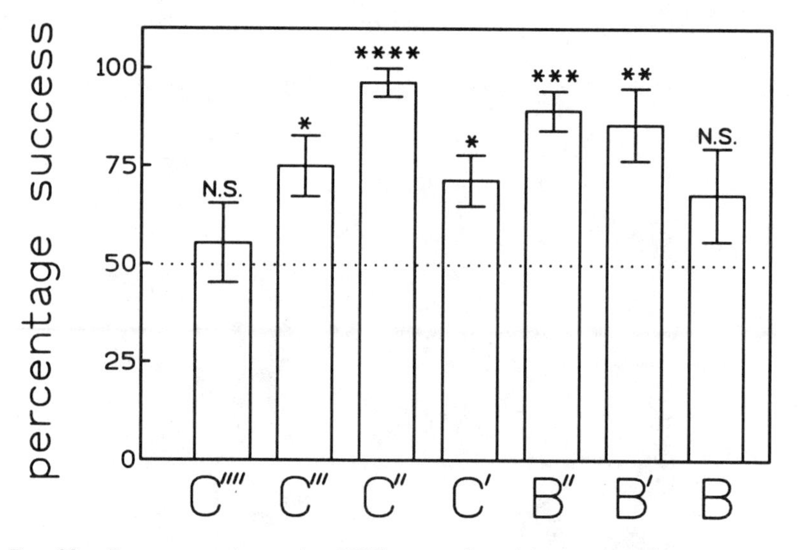

FIG. 23.—Percentage success (± SEM) on probe trials across the seven treatments used in Experiment 13. The dotted line indicates the level of performance expected by chance responding. N.S. = not significant. * $p < .05$. ** $p < .01$. *** $p < .001$. **** $p < .0001$. N.S. = not significant.

The critical results of this experiment can be seen in Figure 23. The most important finding is that the subjects performed at above-chance levels in five of the seven treatments: attending-versus-distracted, looking-over-shoulder, hands-over-eyes, screen-over-face, and buckets (separate one-sample t tests, two tailed, hypothetical mean = 50%; $t[6] \geq 3.240$, $p < .05$ or smaller, in all cases). In sharp contrast, however, one-sample t tests revealed that the subjects performed randomly on both the blindfolds and the new treatment, eyes-open-versus-closed. This absence of comprehension was especially striking in the case of the eyes-open-versus-closed treatment.

A comparison of the obtained results with the alternative predictions outlined in Table 5 reveals that the rule that best predicted the subjects' actual performance was the "face alone" rule: *Gesture toward the person whose face is visible.* This rule correctly predicted how the group would perform in six of the seven treatments, and the one treatment that it failed to predict accurately was mispredicted by the other rules as well.

The ability of this rule to correctly predict how the subjects would perform can be contrasted against the incorrect predictions generated by the other potential rules. For example, if the subjects were using a rule based on the presence of the eyes alone, they should have performed well in the eyes-open-versus-closed treatment. However, they did not. Indeed, as a group, the subjects gestured in front of the person who had his or her eyes open on only 55% of the eyes-open-versus-closed trials (chance = 50%). In addition, if the subjects were basing their choices on the presence of a face with eyes (the "eyes plus face" rule), they should have performed above chance on eyes-open-versus-closed because one of the experimenters matched that stimulus configuration exactly. In contrast, if the subjects had learned only the rule, *Gesture toward the person whose face is visible,* they would respond randomly because both experimenters' faces were visible. This is precisely what the subjects did.

In addition to implicating a particular learned rule, this finding also independently casts doubt on the mentalistic framework because, according to that framework, the subjects should have gestured selectively to the experimenter with his or her eyes open. Call and Tomasello (1994) report that an orangutan that had been enculturated with humans tended to "point" less often when an experimenter's eyes were closed than when they were open. Although suggestive, the effect was not replicated with this subject, nor did a second orangutan (not enculturated with humans) show the same effect.

The subjects' random responding in the blindfold condition is also consistent with the "face alone" rule. If the subjects had learned a rule about eyes, they should have performed correctly on every trial in that treatment. Likewise, a rule about "eyes plus face" could work well here, too, because the experimenter who had the blindfold over his or her mouth had his or her eyes visible as well as most of the face. Only the "face alone" rule accurately predicted that the subjects would not respond above chance because an equal amount of both experimenters' faces was visible: in one the eyes were missing, in the other the mouth (see Fig. 4 above).

The subjects' excellent performance on three of the other five treatments (buckets, screen-over-face, looking-over-shoulder) is consistent with all three rules and therefore cannot be used to implicate one over the other. Nonetheless, their successful performance on these treatments is important because it establishes that, at some point between the early experiments and the present one, the subjects had learned a general rule that cut across the particular probe trial treatments. This is especially salient in the case of the buckets treatment because the subjects had previously received only four trials of this type, and several months earlier at that.

Finally, the slightly lower performance on hands-over-eyes trials was predicted by the "face alone" rule as well (see Table 5). In this treatment, the contrast would be between a part of the face of one experimenter and most of the face of the other, thus suggesting intermediate performance. However, the other rules predicted high performance because, in the case of the "eyes alone" rule, only one experimenter's eyes were visible and, in the case of the "eyes plus face" rule, only one of the experimenters had both eyes and face visible.

The only performance outcome that was not predicted by any of the rules was that obtained in the attending-versus-distracted treatment, in which the subjects gestured at above-chance levels toward the experimenter who was paying attention, despite the fact that all the potential rules had predicted chance performance. However, it is possible that the subjects had learned something specific about this treatment from the previous experiment. Indeed, if the subjects' scores are examined in blocks of two trials across Experiments 12 and 13, they suggest a learning trend: 43%, 64%, 79%, and 71% correct.

In summary, this experiment provided several additional lines of evidence that the subjects were responding at above-chance levels because they had learned rules about particular stimulus configurations associated with reward, not because they possessed a mentalistic understanding of visual perception. In addition, our a priori predictions of which rules would work in which treatments strongly suggest that the subjects' above-chance performances were based on a rule of the type, *Gesture in front of the person whose face is visible.* The results of the eyes-open-versus-closed treatment also revealed that most of the subjects did not yet incorporate eyes as an important feature governing their choices. We say *most of the subjects* because we have some indication from this experiment that one of the subjects, Megan, responded very differently from the others. Across the 28 probe trials in this study, Megan made only a single error. Of even greater interest is that her one error was not in either the blindfolds or the eyes-open-versus-closed treatments. Thus, it was possible that this animal might have learned two rules, one about the face, the other about the eyes. Of course, it is possible that she learned something else entirely, perhaps related to a mentalistic appreciation of seeing (see Chap. VI below).

EXPERIMENT 14

Our final experiment with the young chimpanzees was designed to isolate further the exact structure of the rule system on which the chimpanzees were basing their choices. To this end, we designed three new treatments that each contrasted different stimulus features of the experimenters.

Method

Subjects

The subjects were the same seven chimpanzees used in the previous experiments, and they began the current study the day after completing Experiment 13.

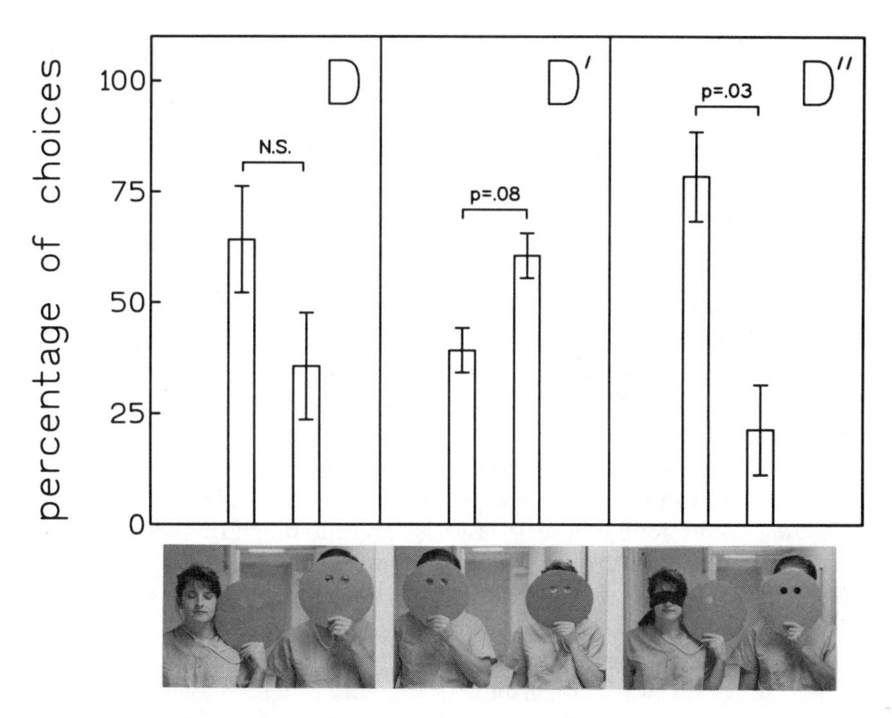

FIG. 24.—Treatments (D, D′, D″) used in Experiment 14 and distribution of subjects' responses (± SEM) to them. N.S. = not significant.

Nature of the Treatments and Predictions

Figure 24 displays three new treatments (D, D′, D″) that were designed to tease apart the hierarchy of rules that the chimpanzees were using for executing their gesturing. In preparation for this study, two of the circular cardboard screens were modified by cutting out large eyeholes so that, if they were held in front of the experimenters' faces, the experimenters' eyes would be clearly visible.

Treatment D (see Fig. 24) contrasted one experimenter who held up a screen next to his or her face, but whose eyes were closed, and a second experimenter who held up a screen in front of her or his face, but whose eyes were open and visible through the holes in the screen. Thus, treatment D contrasted a face with no eyes and eyes with no face. If the subjects were using the "face alone" rule discussed above, they should gesture in front of the experimenter who could not, in fact, see them. In contrast, the "eyes alone" rule (as well as the mentalistic framework) predicted that the chimpanzees would gesture toward the experimenter looking through the screen with eyeholes.

Treatment D′ (see Fig. 24) contrasted both experimenters with the screens covering their faces, but one with eyes open and visible and the other with eyes closed. The "face alone" rule predicted random performance in this treatment; in contrast, the "eyes alone" rule (and the mentalistic framework) predicted a preference for the experimenter with his or her eyes open.

Finally, treatment D'' (see Fig. 24) contrasted two blindfolded experimenters; one of them obscured his or her face as well using a screen, whereas the other held a screen in such a way as not to obscure the face. Thus, the "face alone" rule predicted choosing the blindfolded experimenter with an otherwise unobstructed face, whereas the "eyes alone" rule (as well as the mentalistic framework) predicted random performance.

Treatment D'' differed from all other treatments previously administered to the subjects in that neither of the options facing the chimpanzees was "correct" because neither of the experimenters could see what was happening. Thus, no matter what response the subjects made on these probe trials, they would not be rewarded by the experimenters. In order to ensure that the subjects would continue to respond across the four trials of this type that each received, after the subject gestured on each trial and the experimenter did not respond, the subject was ushered out of the room as usual by the trainer. Once the subject was outside, the trainer gave him or her a small food reward (regardless of the experimenter toward whom he or she had gestured).

Testing Procedure

The subjects were each tested for six sessions of the same structure as those in Experiment 13, with six trials per session, Trials 3 and 6 serving as probe trials, and Trials 1–2 and 4–5 as back-versus-front (C) spacer trials. Every two consecutive sessions contained probe trials of the same type. In order to minimize any possible effect of testing order, the order in which the subjects received the three treatments was counterbalanced as follows. Because there were three different treatments, there was a total of six possible testing orders. These six testing orders were exhaustively and randomly assigned to six of the seven subjects. The remaining subject was randomly assigned to one of the six possible testing orders. All other variables were counterbalanced and randomized following the procedures outlined in previous studies.

Data Analysis

The main data were summarized and analyzed as in previous experiments. In addition, four observers who had not participated in testing the subjects used a handheld stopwatch to record the latency from the time the subject entered the testing unit until he or she responded by gesturing through one of the response holes. After assessing reliability, the four raters' scores were averaged to produce a mean latency to respond for each animal on each of six trial types, the three treatment trial types (D, D′, D″, $N = 4$ trials per animal per phase), and the C trials within each of the treatment phases ($N = 8$ trials per animal per phase). This measure was used to determine in a quantitative fashion how the different treatments affected the animals' response latencies. If the subjects were using a "face alone" rule, we predicted that their hesitations would increase according to the clarity with which one of the alternatives within each treatment easily satisfied the rule. An analysis of Figure 24 indicates the following order for increasing hesitations: D < D″< D′. On the other hand, if the subjects were using an "eyes alone" rule, the ordering should be (D′ = D) < D″.

Results and Discussion

Subjects' Main Responses

The main results of this study are depicted graphically in Figure 24. The bars represent the distribution of the chimpanzees' choices (in percentages) for each of the two stimuli in each of the three treatments. The total percentage of choices for each option in the contrast is displayed above the pictorial representation of each respective option. (As usual, of course, the identity of the experimenters and the side on which they portrayed each option were perfectly counterbalanced across the four probe trials that each subject received of each treatment as well as across subjects.) A planned comparison using Student paired *t* tests revealed that, in treatment D, the group displayed no reliable preference for one of the options over the other, although the trend was in the direction predicted by the "face" rule. For treatment D′, the group approached a statistical preference for the experimenter whose eyes were open behind the screens as opposed to the experimenter whose eyes were closed (Student paired *t* test, $t[6] = 2.121$, $p = .078$). Most striking, however, was the group's clear preference in treatment D″ for the experimenter who could not see because of blindfolds, but most of whose face was visible, over the experimenter who also could not see because of blindfolds, but whose face was, in addition, covered with a screen (Student paired *t* test, $t[6] = 2.828$, $p = .030$).

Latency Data

One of the four raters who scored each trial for the subjects' latency to respond was arbitrarily designated as the main rater for purposes of assessing reliability. The scores of all other raters were compared to this rater and yielded an overall agreement of over 97%. We first analyzed the latency data to determine whether the subjects hesitated most on the probe trials (D, D′, D″) within each treatment phase as compared to the spacer back-versus-front trials within each phase. Because we predicted ahead of time that the latencies would be longer on the probe trials, one-tailed Student t tests were conducted to compare the average latency on the spacer trials with that on the probe trials within each phase. The subjects showed significantly longer latencies on D″ trials than on their surrounding spacer back-versus-front trials (7.35 vs. 2.60 sec, $t[6] = 2.14$, $p < .05$) and approached significantly longer latencies on D and D′ trials than on their surrounding spacer trials (5.40 vs. 2.93 sec and 9.07 vs. 2.67 sec, $p = .06$ and .08, respectively). Having established that on the basis of their response latencies the subjects tended to discriminate the probe trials from the back-versus-front spacer trials, we next examined the data to test the different predictions generated by the "face alone" and the "eyes alone" rule. The results indicated that the average latency to respond on the three treatments was in the order predicted by the "face alone" rule and not that predicted by the "eyes alone" rule (D = 5.40 sec, D″ = 7.35 sec, D′ = 9.07 sec), although a one-way repeated-measures ANOVA indicated that these differences were not statistically significant.

The most striking aspect of these results is that they provide little support for the mentalistic framework. Although it is true that in treatment D′ the subjects did approach a statistical preference for gesturing toward the experimenters when they had their eyes open as opposed to closed behind the screen, this result must be interpreted along with the results from the other treatments as well. In treatment D, when the subjects were confronted with one experimenter who had his or her eyes open and visible behind a screen (just as one of the options in treatment D′) and another experimenter with eyes closed, not only did the subjects not selectively gesture toward the one who could see, but if anything they tended to prefer the one who could not see but whose face was unobstructed. In addition, in treatment D″, the subjects did not respond randomly as predicted by both the "eyes alone" rule and the mentalistic framework but instead showed a clear preference for selecting the experimenter who could not see but whose face was mostly visible. Finally, the examination of the latency data indicated that the subjects tended to hesitate least when the "face alone" rule could be easily satisfied and most when it could not.

We interpret this pattern of results as indicating that most of the subjects had learned (or were in the process of learning) a hierarchy of rules about the face and eyes. The most robust of these rules appeared to be, *Gesture toward the person whose face is visible,* and this rule appeared to guide their behavior irrespective of whether eyes were present. However, the results of treatment D′ suggest that, when no face was available to guide their choices, the subjects might be able to rely on eyes as a secondary cue.

GENERAL DISCUSSION OF EXPERIMENTS 7–14

We undertook the experiments reported in this chapter with the aim of testing certain procedural explanations for why the subjects had not performed according to the mentalistic framework in the initial set of experiments reported in Chapter III. However, the eight experiments reported in this chapter provided little additional evidence to support the mentalistic framework's account of what young chimpanzees know about seeing. Indeed, the overall pattern of the results from Experiments 5–11 is consistent with the view that the subjects gradually learned to respond to certain features of the two experimenters. The results from Experiments 12–14 further suggest that the most important of these features was the experimenters' faces and that when that cue was not available they might also be able to rely on the presence of eyes.

Regardless of the exact nature of the stimuli controlling the chimpanzees' behavior, we obtained no compelling evidence that the chimpanzees appreciated the special significance that eyes had in subjectively linking one of the experimenters to them. Perhaps the best evidence in favor of this rather skeptical view is that in Experiments 12–14 the subjects performed randomly on their initial confrontation with each new treatment in which simple rules concerning the stimulus feature of eyes or faces could not work. This was true for attending-versus-distracted (Experiment 12), eyes-open-versus-closed (Experiment 13), and treatment D and to a lesser extent D′ (Experiment 14). This initial diagnosis of comprehension is important because, as the results of the attending-versus-distracted probe trials in Experiment 12 versus 13 clearly show, the subjects were primed for rapid learning.

The final aspect of the results of these studies worthy of comment here is that they demonstrate in a rather dramatic fashion how behaviors seemingly indicative of a mentalistic appreciation of visual perception may, in fact, have nothing whatsoever to do with such a phenomenon. The chimpanzees' automatic and robust tracking of the distracted experimenter's line of sight in Experiment 12, with no comparable apparent comprehension of the importance of this posture for the experimenter's subjective attentional focus, indicates how the use of the eyes of others in no way uniquely implies a corresponding view into their mind. We assess the strengths and weaknesses of this position further in Chapter VI.

V. ASSESSING VALIDITY WITH YOUNG CHILDREN

In Chapters III and IV, we described the results of nonverbal tests to determine whether chimpanzees interpret vision as a mental process. Although we have explained how the two alternative explanatory frameworks logically generated different predictions about how the subjects should have performed on each of the tests, logic alone cannot guarantee their validity. That is, no matter how intuitive our nonverbal designs appear to be as measures of whether chimpanzees interpret vision as both a behavioral and a mental event or strictly as a behavioral event, additional data are desirable as independent support for the view that the tasks presented to the chimpanzees do indeed measure what we think they measure. One method of doing so is to administer the same tests to young children (see Köhler, 1927; Povinelli, 1993; Premack & Dasser, 1991). In this fashion, an appropriate theory of the task can be constructed (Cole & Means, 1981).[5]

The logic behind using research with young children as a means of assessing task validity is that such research provides a point of departure. Evolutionary reconstructions of morphological traits explicitly ask which identifiable features of given lineages are derived and which are primitive. By analogy, in the case at hand, we are interested in determining whether

[5] An additional problem concerns how to cope with the cases in which the verbal and nonverbal measures suggest different interpretations of what young children understand at a given age. Some may not see this as a large problem because at some point the two measures will converge to produce a strong diagnosis of the presence of an understanding or a particular theory of mind concept. Yet, ironically, some developmental researchers have expressed envy over the fact that researchers working with animals are not plagued with the problem of interpreting what their subjects mean when they provide a particular linguistic utterance (see Chandler et al., 1989). This raises the ugly problem of determining how we prioritize different results from verbal and nonverbal behavior. In the cases where both measures point to a similar age transition, there is little conceptual problem. Where the measures point to different ages, we can offer no better solution than to suggest that perhaps ultimately, with a complete understanding of the ways in which children of differing ages understand mental state terms and a complete understanding of the theory of nonverbal tasks, a union of verbal and nonverbal measures will be achieved.

a mentalistic understanding of visual perception is a shared feature of chimpanzee and human psychological ontogeny or whether it represents an autapomorphic (uniquely derived) feature of human ontogeny. With this question firmly in mind, consider one type of skeptic who might doubt the validity of our largely negative results with the chimpanzees. Such a skeptic could argue that our task is really measuring more than a mere understanding of the cognitive connection engendered by visual perception. For instance, the skeptic's interpretation of the task could be that the chimpanzees need to understand the seeing-knowing relation in order successfully to gesture toward the person who can see them.

From this perspective, success on our tests could depend on an organism's ability to reason as follows: *That one can see me; therefore, if I gesture toward her, she will know I want food;* or conversely: *That one cannot see me; therefore, he will not know that I have gestured toward him.* This interpretation could be used to argue that young chimpanzees do, in fact, understand vision as engendering a mental connection to the world but perform randomly on our tests because they are really measuring an aspect of psychological understanding that emerges later in human development (i.e., an aspect that is more "sophisticated"). If correct, our task would demonstrate that young chimpanzees do not attend to the relation between seeing and knowing, not that they fail to understand the more general psychological connection engendered by seeing. The potential plausibility of such an account is bolstered by the fact that, in a previous study with five of the seven chimpanzees used in the studies reported in Chapters III and IV, none of the subjects showed an appreciation of the connection between seeing and knowing (see Povinelli et al., 1994; separate research using the same task has demonstrated that young children do not typically perform well until they are 4 years of age; see Povinelli & deBlois, 1992).

From the standpoint of assessing the task's validity, the high-level theory of our task outlined above generates at least one very clear comparative prediction: 2- and 3-year-old human children ought to perform quite poorly on the tests reported in Chapters III and IV, whereas older 4-year-olds ought to perform quite well. This prediction can be neatly derived from research by John Flavell and others (reviewed in Chap. II) that young 3-year-olds have great difficulty with tasks that require understanding the mental *consequences* of visual perception yet have no difficulty with tasks that require understanding the fact that visual perception subjectively connects people to the world (Flavell, 1988). Thus, if our hypothetical skeptic's explanation of the chimpanzees' trouble with our tests is correct, children between the ages of 2½ and 3½ should have great trouble also. On the other hand, if the task is measuring the earlier emerging understanding of seeing, then 2½–3½-year-olds should perform quite well.

Note that the exact approach outlined above is dictated because of the negative evidence that we obtained. Had the chimpanzees performed extremely well on all variants of our tests, an assessment of task validity would have proceeded by examining a very different set of predictions about young children's performance. In that case, we would have been forced to make predictions about the *youngest* age at which children could succeed on the tests. Determining the youngest age at which children would perform according to the mentalistic framework would have been necessary because in that case the most salient skeptics would arise from the behaviorist side of the aisle. These skeptics would contend that the chimpanzees' success was not based on a mentalistic understanding of vision. In this case, the task's validity as a measure of the "aboutness" (or referential property) of seeing would hinge on the prediction that children younger than about 2½ years should perform poorly, as this is the youngest age at which clear evidence of understanding the subjective aspect of seeing has been demonstrated (Lempers et al., 1977). On the other hand, no definitive minimum age has been demonstrated for passing level 1 perspective-taking tasks.

EXPERIMENT 15

In this experiment, we tested several age ranges of preschoolers on a test as analogous as possible to the one used with the chimpanzees. First, we trained and tested preschool children across a several week period to gesture toward the experimenters in a situation comparable to that experienced by the chimpanzees. Second, we selected several of the visual deprivation treatments used with the chimpanzees and tested the children using the same probe trial technique.

Method

Subjects

The subjects were 47 children from one of four preschools in the Lafayette, Louisiana, area whose parents had granted permission for them to participate in the research project. At the time of testing, the children were divided into four age groups: from 2-7 to 3-1 ($N = 11$, $M = 2\text{-}11$, five boys, six girls), from 3-2 to 3-9 ($N = 12$, $M = 3\text{-}5$, eight boys, four girls), from 3-10 to 4-5 ($N = 12$, $M = 4\text{-}1$, seven boys, five girls), and from 4-6 to 5-1 ($N = 12$, $M = 4\text{-}9$, four boys, eight girls). The youngest two age groups that we tested have previously (and repeatedly) been shown to perform poorly on tasks that require them to understand the mental consequences of visual perception but to perform well on tasks that require them to understand only the subjective connection engendered by such perception (see Chap. II). Each child participated in the study once a week over a period of 6–8 weeks. All subjects were members of working- and middle-class families.

Training Procedure

The subjects were initially trained across a minimum of three and a maximum of six sessions. At the start of the first session, individual children were brought into a quiet, familiar room at the day care by two adult experimenters with whom the children were familiar. The room contained a long table. The children were shown two sheets of construction paper, each with an outline of a hand traced on its surface (one was the outline of a left hand, the other that of a right hand). The children were asked to identify the tracings, and, after they answered correctly that they were hands, one of the experimenters showed the children how the handprints could be placed at opposite ends of the table (the distance between them was approximately 1 m). Next, the children were told that they were going to play a game in which one experimenter (the target experimenter) would kneel behind the table in front of either the handprint on the left or the one on the right. The children were told that the target experimenter would give them a sticker if they placed their hand on the handprint directly in front of them.[6] Next, the children were administered one to three practice examples of placing their hand on the prints in front of the target experimenter.

Each training session consisted of nine trials. At the start of each trial, one of the experimenters led the children to a neutral starting position approximately 1.5 m from the table and instructed them to turn away from the table and face the wall. On Trials 1–2, 4–5, and 7–8, the target experimenter knelt behind the table in front of one of the handprints so that his or her upper torso and head were completely visible and approximately eye level with the children. The target experimenter fixed his or her gaze on the handprint directly in front of him or her. Thus, as with the chimpanzee testing, the experimenter did not make eye contact with the children. The other experimenter, who was waiting with the children, told them that, as

[6] In pilot testing, we attempted to teach children to point in the same manner as the chimpanzees, but most children were extremely reluctant or embarrassed to come forward and point at the experimenters. In addition, the points that they did produce were often so subdued that it was difficult to determine exactly to whom they were referring. In contrast, the children rapidly adapted to the procedure of extending their arm and placing it on the handprint in front of the target experimenter in order to obtain their rewards (stickers).

soon as they heard a knock, they should turn around and go and put their hand down to get the sticker. The target experimenter then quickly knocked under the table midway between the two handprints. The children then turned, approached the table, placed their hand correctly, and were rewarded with a sticker, which they then placed on a piece of paper. On Trials 1–2, many of the children had to be prompted to put their hand on the paper after they approached the table but then simply stood in front of the experimenter, waiting to obtain a reward.

Trials 3, 6, and 9 were probe trials comparable to the baseline (A) probe trials used with the chimpanzees. On these trials, both experimenters positioned themselves behind the table in front of a handprint while the child faced the wall. One held out an empty hand; the other held out his or her hand in the same way, but it contained a sticker. Both experimenters fixed their gazes on the handprints. One of the experimenters knocked midway between the prints, and the child turned around and responded. Most children approached the table and placed their hand on one side or the other. Several approached the experimenter offering the sticker and initially reached directly for the sticker. If this occurred, the experimenter closed his or her hand in order to prevent the child from taking the sticker while simultaneously maintaining his or her gaze on the handprint outline. Several subjects corrected themselves automatically by placing their hands on the handprint; others had to be prompted during the first session by asking, "Where do you put your hand if you want the sticker?" The purpose of this training was to encourage the subjects to look back and forth between the two experimenters (at least their hands) before they made their choice. This was done so that on the critical probe trials that were to follow, in which there were no offering gestures by either experimenter, the subjects might shift to the experimenters' faces before selecting one of them.

Four training schedules were constructed that counterbalanced or randomized the key variables of the procedure, including the identity of the experimenter who served as the target on Trials 1–2, 4–5, and 7–8, the side of the correct choice, and the experimenter who was correct on the probe trials. On the three probe trials within each session, one side and one experimenter were correct once, and the other side and the other experimenter were correct twice. These two variables were approximately counterbalanced across training sessions to ensure that no experimenter or side was correct a disproportionate number of times. The children were randomly assigned to training schedules (without replacement) across sessions. Each subject was tested for a minimum of three sessions with the criterion of two sessions in a row in which they performed correctly on all spacer and probe trials. Children were typically administered one training session per week, but occasionally some children received two sessions per week. Once the subjects had met criterion, they were advanced to the test phase. No child took more than 6 weeks to complete the training phase (M = 3.4 weeks).

Testing Procedure

The children were tested on three examples of one experimenter "seeing" and the other "not seeing" that were identical to several that we had used with the chimpanzees: back-versus-front (C), hands-over-eyes (C′), and screen-over-face (B″). The stimulus configuration and materials were the same as those used with the chimpanzees. On probe trials (as on the other trials), the experimenter who could see gazed at the handprint in front of him or her. Neither experimenter held out his or her hands on these probe trials. The box containing the rewards (stickers) was placed midway between the two experimenters, as the food had been in the chimpanzee studies. If the child placed a hand on the handprint in front of the experimenter who could see, he or she was rewarded by that experimenter; if not, he or she was not rewarded but was told to try again (at which point the preparation for the next scheduled trial began).

After advancing from training, each child was administered three test sessions of nine trials each (one test session containing three probe trials of one of the treatment types). Thus, each subject received one session in which the probe trials were of back-versus-front, one session of hands-over-eyes, and one session of screen-over-face. Each of the sessions was structured in the same general fashion as the training sessions: Trials 1–2, 4–5, and 7–8 serving as spacing trials with one experimenter behind the table looking at the handprint and Trials 3, 6, and 9 designated as probe trials with both experimenters present. Three test schedules were constructed that approximately counterbalanced all the variables described for the training schedules.

In order to control for possible practice effects, the children in each age group were randomly assigned to the three test session types using an all-possible-orders design. Thus, each of the three treatment types occurred equally often in the first, second, and final session across the subjects in each age group. (A minor exception to the assignment existed in the youngest age group, which was composed of 11 subjects instead of 12. For this age group, one schedule remained unassigned.) Finally, the three test treatments were randomly assigned to the three test schedules so that each schedule appeared equally often in each session position and equally often for each treatment type. Typically, one session was administered per week, although on some occasions the children received two sessions per week. In order to minimize learning across sessions, we avoided administering two sessions immediately following each other. Because we used an exhaustive all-possible-orders design, the data could be analyzed by treatment even if the subjects were to show a learning effect across test sessions.

Results

Training

Thirteen children began and/or completed the training phase but were dropped from the study for several reasons. Seven were dropped because they left their preschool during or immediately after completing the training phase. Two additional subjects were lost because they left their preschool after completing training and their first test session. Two other children (both from the 3-11 to 4-5 age group) were dropped because they did not achieve the training criterion within the required number of sessions. Two more children were dropped from the oldest group—one because of a testing error on the part of the experimenters, the other because he or she aged out of the assigned group before testing began. All these subjects were replaced with new children.

The important result of the training is that by the end of training the subjects all were clearly comparing the two experimenters before making their choice. The results of the probe trials from the training sessions are presented in Figure 25. The graph depicts the grand mean of the children's average score (zero to three correct for each child) in each age group across sessions. Only the first three sessions are shown. Two subjects in each of the three oldest age groups required a fourth training session to meet criterion. Seven children from the youngest age group required more than three sessions (range = four to six sessions). As can be seen, all age groups showed a similar learning curve across the training sessions to criterion. However, although the oldest group showed a trend toward somewhat more rapid learning and the youngest group showed the least rapid learning, a 4 (age) × 3 (session) ANOVA indicated no effect of age but did reveal a significant effect of session ($F[2, 86] = 22.95$, $p < .0001$). There was no interaction between the two. Thus, all subjects showed a gradual improvement across sessions on the probe trials where they chose between the two experimenters, one holding out a sticker, the other holding out an empty hand.

Curiously, on Trial 1 of Session 1, only the 3-2 age group approached an above-chance performance (binomial test, nine successes, three failures, $p = .054$). All other age groups' Trial 1 performance did not differ from chance (58%, 67%, and 72% correct for the 2-7, 3-11, and 4-6 age groups, respectively). However, if the data for the first training session (three probe trials) as a whole are examined, all the groups performed above chance. One-sample t tests (two tailed, hypothetical mean = 1.5 responses correct by chance) revealed that the youngest age group as well as the two oldest

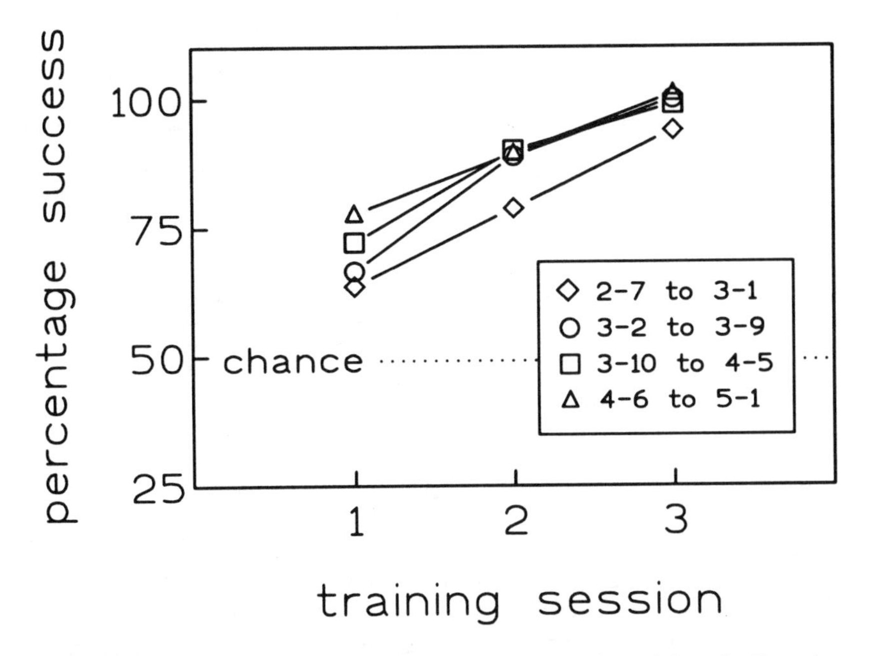

FIG. 25.—Results of training sessions by age with preschool children for Experiment 15. Only the first three sessions are shown as only six of 36 children required a fourth session to meet the training criterion (for additional details, see the text).

age groups performed above chance in Session 1 ($t[10] = 2.52$, $p = .031$, and $t[11] = 3.22$ and 4.43, $p = .008$ and $.001$, for the youngest group and the two oldest groups, respectively) and that the 3-2 age group approached an above-chance performance ($t[11] = 2.03$, $p = .067$).

Testing: Gender Effects

In order to examine the results for gender effects, the subjects were divided into boys ($N = 24$) and girls ($N = 23$), and each subject's three test sessions were averaged to yield a single score. A Student *t* test indicated no significant difference in performance between boys and girls (girls' $M = 85.0\%$ correct, boys' $M = 82.4\%$ correct). Given that there was no gender effect, all subsequent analyses reflect data collapsed across this factor.

Testing: Main Results

Initially, we examined the data to test our main hypothesis that all age groups would perform significantly above chance. To test this hypothesis,

TABLE 6

YOUNG CHILDREN'S AVERAGE PERCENTAGE CORRECT BY CONDITION AS A FUNCTION OF AGE, EXPERIMENT 15

		TREATMENTS		
AGE GROUP	N	Hands-over-Face	Screen-over-Face	Back-versus-Front
2-7 to 3-1	11	69.7*	51.5[a]	75.8*
3-2 to 3-9	12	75.0*	83.3**	94.4****
3-10 to 4-5	12	83.3***	83.3***	100[b]
4-6 to 5-1	12	91.6****	94.4****	94.4****

NOTE.—Data were analyzed using one-sample t tests (two tailed, $df = 11$, hypothetical mean [chance performance] set at 50%). All t's ≥ 2.28 and ≤ 11.86.

[a] N.S.

[b] A t test was not possible for this cell because group variance was equal to zero.

* $p < .05$.

** $p < .01$.

*** $p < .001$.

**** $p < .0001$.

we examined each age group and treatment separately by averaging the number of probe trials correct (zero to three) for each subject for each treatment session and then obtaining group means for each age group–treatment type combination. One-sample t tests (two tailed, hypothetical mean = 1.5 trials correct) were conducted for each of the treatment type–age group results. The results of these analyses are presented in Table 6 and clearly indicate that, across age groups and treatments, the children performed at levels above what would be expected had they been responding at random, except for the youngest age group, whose performance on the screen-over-face treatment did not differ from that expected by chance.

Next, we examined our secondary hypothesis that the youngest two age groups of children should perform equally well as the oldest children. A 4 (age) × 3 (treatment) ANOVA indicated effects of age ($F[3, 43] = 5.24$, $p = .0036$) and treatment ($F[2, 86] = 6.43$, $p = .0025$). There was also a marginally significant interaction effect between the two ($F[6, 86] = 2.12$, $p = .0593$). Tukey post hoc comparisons for the age effect indicated that the youngest group performed significantly more poorly ($p < .01$) than the three oldest groups (64% vs. 83%, 89%, and 93% correct, respectively), which did not differ from each other. Post hoc comparisons for the treatment effect indicated that it was due to the fact that the subjects performed significantly better in the back-versus-front treatment (90.5% correct) than in the screen-over-face treatment (76.8% correct) ($p < .01$). The results for hands-over-eyes were intermediate (79.9% correct) and did not differ significantly from either of the other two treatments. A statistical decomposition of the simple main effects of the marginally significant age by treat-

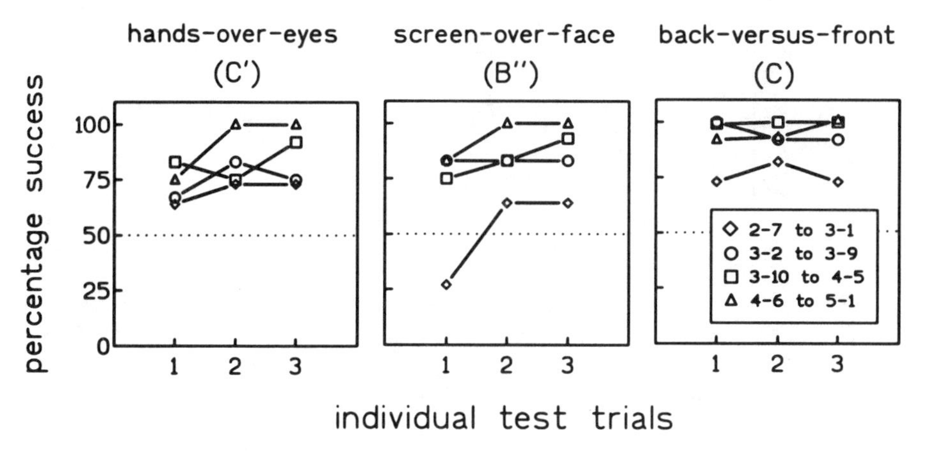

FIG. 26.—Mean trial-by-trial probe trial results separated by treatment and age group for Experiment 15.

ment interaction showed that the effects of age at screens and treatment type for the youngest age group contributed significantly to the interaction ($F[3, 97] = 8.27$, $p < .0001$, and $F[2, 86] = 7.13$, $p < .001$, respectively). Both of these contributions to the interaction were due to the youngest age group's significantly poorer performance on the screen-over-face treatment when compared with the other age group–treatment type combinations.

The data were also examined for a learning effect across sessions independent of treatment. A 4 (age) × 3 (session) ANOVA revealed no effect of session but did reveal an effect of age ($F[3, 43] = 4.78$, $p = .0058$). There was no interaction between the two. Tukey post hoc comparisons indicated that the age effect was due to the fact that the oldest age group performed better than the youngest group ($p < .01$). Performance for the middle two age groups was intermediate and did not differ from either the oldest or the youngest age groups. Finally, we examined the data on a trial-by-trial basis to determine whether the subjects showed within-session learning effects. This was important to determine because, in order to conclude that they understood the treatments immediately, we needed to demonstrate that, independent of treatment, the children responded correctly on their very first trial.

In order to examine this possibility, we conducted two separate analyses. First, we examined each age group's performance on a trial-by-trial basis within each treatment type. The results of this analysis are depicted in Figure 26. The pattern of these results indicates that, although for the hands-over-eyes and screen-over-face treatments (but not the back-versus-front treatment) there appears to have been some within-session learning, the Trial 1 performance in all treatments for all four age groups appears

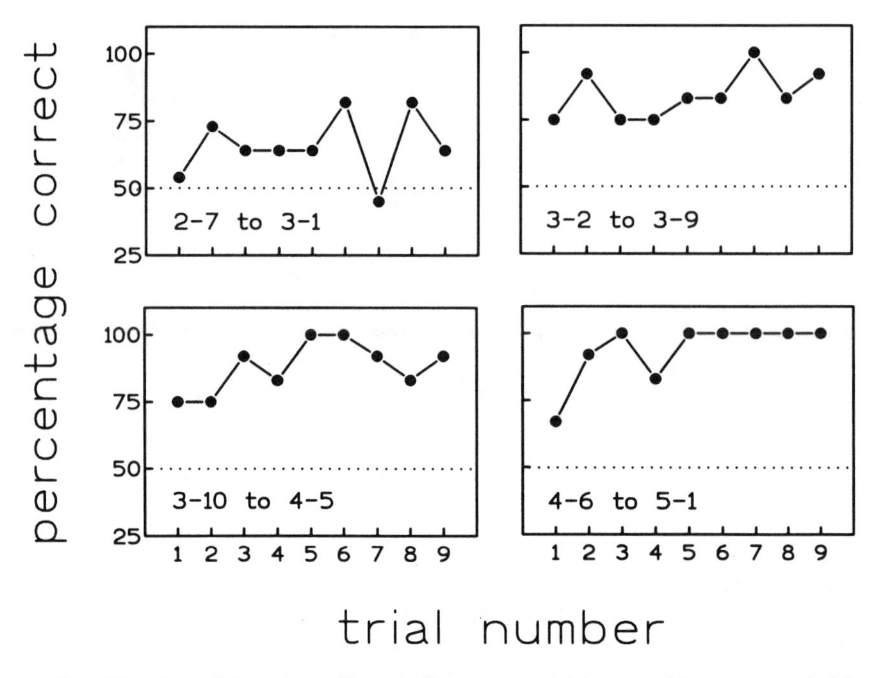

FIG. 27.—Overall learning effects in Experiment 15 by age group as revealed by mean trial-by-trial probe trial results in order received irrespective of treatment.

to have been better than that predicted by chance responding, except in screen-over-face for the youngest group. We therefore combined the Trial 1 data for all age groups and compared the subjects' performance to that predicted by chance. Trial 1 data for each condition well exceeded that expected by chance (binomial test, $p < .01$ in all cases): hands-over-eyes, 72% (34 of 47 children) correct; back-versus-front, 91% (43 of 47 children) correct; screen-over-face, 68% (32 of 47 children) correct.

For the second trial-by-trial analysis, we examined the subjects' trial-by-trial data in the actual order in which the probe trials were received, irrespective of treatment. These results are shown in Figure 27. The pattern of results from this analysis could reveal whether the subjects performed above chance on their very first test trial, regardless of what treatment they received. All children combined were 68% (32 of 47 children) correct on Trial 1, a performance that differs from chance (binomial test, $p < .01$). However, as Figure 27 reveals, the youngest age group's Trial 1 performance was not robust (only 54% of these children were correct on Trial 1). These two analyses allow us to conclude with some certainty that, before any within- or between-session learning could possibly have occurred, as a group the subjects responded to the treatments in the manner predicted by the mentalistic framework. The only exception to this appeared to be in the youngest age group.

Discussion

In general, the results of this study support the prediction generated by the theory of the task that interprets it as a measure of an organism's understanding of seeing as attention, not as a measure of an understanding of seeing as a knowledge acquisition device. Even older 2-year-olds performed quite well on most of the tasks, indicating that a robust understanding of the relation between perception and knowledge is not required to gesture toward someone who can see you as opposed to someone who is visually disengaged from the situation. For those who feel that such an outcome was never in doubt, it is useful to consider the results of previous attempts to validate nonverbal tasks as measures of levels of understanding visual perception. Povinelli and his colleagues designed and implemented a nonverbal test for studying the seeing-knowing relation in chimpanzees and rhesus monkeys (Povinelli et al., 1990; Povinelli et al., 1991). However, although from an intuitive perspective the task seemed to be a good measure of an organism's understanding of the link between visual perception and knowledge formation, some researchers argued that perhaps the chimpanzees were solving the task using other cues. Povinelli and deBlois (1992) noted that a straightforward way of testing this interpretation would be to administer the same task to young 3- and 4-year-olds. If the task were measuring the seeing-knowing relation, the 3-year-olds should perform at chance levels, whereas the 4-year-olds should do quite well. Each child was administered 10 trials, and in this case the results fit the prediction of the high-level interpretation of the task. Povinelli and deBlois concluded that the task was sensitive to an organism's understanding of the seeing-knowing relation (at least when limited to very few trials). However, because the chimpanzees tested by Povinelli et al. (1990) were given so many repeated trials, interpreting their performance (including their transfer test) was still problematic (see Povinelli & deBlois, 1992).

Additional evidence in support of the lower-level theory of the task comes from our informal questioning of many of the subjects at the end of their final test session. The children were taken to the front of the table midway between the two experimenters, shown a particular treatment, such as screen-over-face, and then asked, "Where would you put your hand if we were like this?" Virtually all the children performed correctly. They were then asked, "Why would you point to her/him and not me?" Additional questions were asked in a variety of formats depending on how the individual children answered. The pattern of the younger children's replies was remarkably consistent. They typically made reference to the eyes ("Your eyes aren't there," "her eyes") or seeing ("You can't see me"). When further questioned (in a number of ways) to see whether they appreciated that the experimenter who could see them knew where they had gestured whereas the other did not know, none of the younger children gave any indication that they understood this fact. However, several of the oldest children made reference to the epistemological significance of the fact that one of the experimenters could see them (e.g., "If I put my hand there, you wouldn't see me, so you wouldn't even know I had done it!").

Several of the specific results of this experiment are of further interest. First, it is curious that across the two species (chimpanzees and humans) a similar treatment effect was obtained in the back-versus-front condition. In both species, this was by far the easiest probe trial type. The chimpanzees averaged 92% correct on their first two trials of this treatment, and five of six animals were correct on Trial 1; the 47 children from all four age groups combined were 91% correct on Trial 1. From one perspective, it is easy to explain why this treatment was easier for the subjects of both species: the answer was more obvious. However, at another level, the exact source of this clarity becomes obscure. As we noted in discussing Experiment 1, it is unclear whether the back-versus-front treatment was easier because of the prior training history that the subjects had received in the context of our studies or whether the frontal stimulus had a valence even before explicit training in our test settings began. We suspect the latter because, in pilot testing with experimentally naive 3-year-old chimpanzees at the New Iberia Research Center, we obtained preliminary evidence that they spontaneously used their begging gestures more when someone stood facing their cage than when he or she stood in the same position facing away (for additional naturalistic evidence that the front/back distinction is one that chimpanzees learn on their own, see Tomasello et al., 1994). At any rate, regardless of when this valence developed, it is still not clear from such data alone whether the attraction was based on (*a*) an understanding of the subjective connection engendered by seeing, (*b*) learned rules of social responding, or (*c*) an understanding of attentional connection without an explicit understanding of the role that eyes play (see Baldwin & Moses, 1994). Thus far, we have no coherent reason to exclude the latter.

The youngest children's poor performance during the screen-over-face treatment is also of interest. On the one hand, given that in general the youngest age group made the most errors, it is possible that their understanding of the intentional aspects of seeing may have been more fragile than that of the older group. Still, the subjects who did make errors in this treatment (as well as in hands-over-eyes) often did so in a remarkably attentive fashion. That is, they proceeded to the incorrect option (the person who could not see them), placed their hand down, and waited for the person to give them a sticker. Unfortunately, the videotapes of the subjects are not of sufficient quality to retrieve information on exactly where they were looking. Nonetheless, if the general attentional focus of others is more salient to young children than the specific role that the perceptual senses play in generating and maintaining that focus, then it is curious that the youngest children did worst in the treatment that provided the clearest elimination of the experimenter's attentional focus. More curious still, an analysis of the level 1 visual perspective tasks administered by Lempers et al. (1977) suggests that children even younger than our youngest group could have easily answered the question, "Can she see your hand right now?" in the context of the screen-over-face treatment. They would also know to remove the screen from in front of the face of the experimenter if we asked them, "Make it so that she can see your hand."

If children as young as 2–2½ years of age can perform well on such level 1 tasks, why did they not do better on our tasks? First, Lempers et al. (1977) allowed readministration of each task in order to determine what the subjects' highest level of competence might be. Thus, in many cases, it is difficult to compare the Trial 1 performances of our youngest children to theirs. Indeed, in a recent study, Gopnik, Melzhoff, and Esterly (1995) reported data consistent with an emergence of level 1 understanding of seeing in children of about 30 months or so, but not younger. Second, the Lempers et al. (1977) data set also reveals that 2–2½-year-olds had much greater difficulty when the task involved one person seeing some aspect of themselves (tasks 5, 13, and 14). For example, when subjects were asked to place their hands so that an observer could not see them, very few of the youngest children were able to do so. Third, it is possible that verbally reporting what another can see or actively placing an obstacle between an observer and an object to deprive that individual of "seeing" ("percept diagnosis" and "deprivation," as Lempers et al. describe them) may not require a understanding of the "aboutness" of perception and instead may be solved by geometric calculations of line of sight (Baron-Cohen, 1994). Finally, of course, it is quite possible that the Lempers et al. tasks are simply more sensitive measures of the early understanding of seeing as attention. We discuss the implications of these possibilities further in the next chapter.

VI. CONCLUSIONS

Before we begin to outline our principal conclusions from the 15 experiments reported in the previous three chapters, it will be useful briefly to redescribe the two theoretical frameworks that we set out to test. The "mentalistic" framework portrays chimpanzees as possessing some kind of a folk theory of seeing. We have not described the exact nature and scope of that theory but rather sought to identify some of the minimal components that such a theory would possess. One indispensable element of this kind of a folk theory would be that chimpanzees appreciate that visual perception subjectively connects organisms to the external world (see Flavell, 1988). In this sense, in order to qualify as a genuinely mentalistic understanding of seeing, chimpanzees would have to appreciate that seeing refers to or is "about" something—in other words, they must interpret seeing as an intentional event. Conversely, the "behaviorist" framework argues that chimpanzees possess or develop algorithms for processing information about the visual gaze of other organisms but that these algorithms are not intentionally grounded; the chimpanzees do not appreciate that seeing is "about" anything. This would be consistent with the possibility that they do not appreciate others as intentional agents (see Povinelli, 1993; Tomasello, Kruger, & Ratner, 1993). Notice that there remains much unspecified about what a chimpanzee's account of seeing would be under both of the frameworks. We shall provide additional details as we proceed.

EVIDENCE FOR A MENTALISTIC UNDERSTANDING OF SEEING IN YOUNG CHIMPANZEES

For those who wish to hold out for an intentional understanding of visual perception on the part of our young chimpanzee subjects, there are three main aspects of our results that could be used to bolster such a view. First, the subjects did display an immediate disposition to gesture toward the experimenter who could see them in the naturalistic treatment of back-versus-front. In Experiment 1, five of six subjects gestured toward the experimenter facing forward on their very first trial with back-versus-front, and, on Trial 2, all six subjects gestured appropriately. This performance remained high and stable across the experiments. Thus, one could argue that, in the treatment that most resembled that which the chimpanzees would encounter in their normal day-to-day interactions, the subjects immediately performed according to the mentalistic framework.

Although such an argument is possible, it has (in our minds) three critical weaknesses. First, it ignores the fact that in the back-versus-front treatment one of the experimenters was configured in exactly the same manner as the experimenters had been during their lengthy training history, thus providing ample room for the behaviorist framework to predict that the subjects should gesture toward the person facing forward. The results of Experiment 3 (see Fig. 8 above) provided fairly compelling evidence to support this interpretation in that the subjects responded at chance in the initial set of looking-over-shoulder trials. Second, using success on the back-versus-front treatment as evidence to support the mentalistic framework also ignores the fact that learning that occurred outside the testing context could have generalized to the test setting. Thus, if it is true that the back-versus-front configuration is something naturally experienced by the subjects every day, then there is no reason to use correct performance in this treatment as unique evidence for a mentalistic understanding of visual perception. The chimpanzees could easily have learned appropriate rules (based solely on observable contingencies) that would lead them to gesture in front of the experimenter facing forward (see Tomasello et al., 1994). Finally, at best such an argument could be used to support the view that chimpanzees might have some conception of *attention,* but we see no unique evidence that such a conception uniquely incorporates visual perception as a mediating mechanism.

The second aspect of our results that could be viewed as consistent with the mentalistic framework is that the subjects showed an apparently rapid acquisition curve on a second naturalistic treatment, looking-over-shoulder. In fact, two of the subjects, Jadine and Candy, gestured toward the experimenter who was looking over his or her shoulder appropriately from the outset (see the "Discussion" section of Experiment 10). As a group, the subjects were above chance by the second set of four probe trials that they were administered. An argument could be constructed around these results that would maintain that such a rapid acquisition curve demonstrates that the subjects appreciated that they were connected in a subjective fashion to the experimenter who was looking.

Again, however, we are cautious of such an interpretation. First, there is no evidence from their first and second pair of probe trials that they selectively gestured toward the experimenter who was looking over his or her shoulder (see Fig. 8). Thus, unlike the data for back-versus-front, the subjects' performance was not above chance during their first and second encounter with these probe trials. Second, the results of Experiments 12–14 support the interpretation that the subjects had learned a broader rule about the visibility of the face (and perhaps the eyes) during the course of Experiments 1–9. If true, this would mean that the consistently above-chance performance of the subjects on the looking-over-shoulder treatment in Experiments 10 and 11 was the result of learning, not the attribution of the attentional significance of seeing. Finally, Jadine's performance could be seen as an outlier—just as Brandy performed at 25% below chance in Experiment 3, Jadine may have randomly performed at 25% above chance. Candy, on the other hand, was 75% correct from the outset without having received anywhere near the number of probe trials as Jadine and the others. However, it may be that she learned the general rules faster because she received all her test trials in the configuration in which the experimenters sat near the testing unit.

The third aspect of our results that is relevant here is the experimental demonstration of gaze following or joint visual attention in these young chimpanzees. Some researchers see gaze following in human infants as evidence of their intentional understanding of seeing (e.g., Baron-Cohen, 1994). However, for reasons outlined in Chapter III, we question this interpretation (see the "Discussion" section of Experiment 12).

EVIDENCE AGAINST A MENTALISTIC UNDERSTANDING OF SEEING IN YOUNG CHIMPANZEES

We believe that the best interpretation of the results is that our young chimpanzees displayed no clear evidence of appreciating the mental connection engendered by visual perception. Below, we list our principal reasons for coming to this conclusion.

1. In none of the three object conditions (buckets, blindfolds, screen-over-face) did the subjects display evidence of an immediate disposition to gesture toward the person who was visually connected to the situation. This was true despite our extended efforts to rule out motivational and associationist factors that might have interfered with an underlying mentalistic understanding of seeing. In addition, the subjects were given extended exposure to all the objects before they were used in the tests, and, perhaps more important, the subjects had access to a variety of objects with similar properties since birth.

2. In four of five naturalistic treatments (hands-over-eyes, looking-over-shoulder, attending-versus-distracted, eyes-open-versus-closed), the subjects

showed no immediate appreciation of the significance that only one of the experimenters could see them (let alone that they were paying attention).

3. As a behaviorist account would predict, the subjects showed a learning curve from Experiment 1 to Experiment 13. For example, a meta-analysis of the subjects' performance on the screen-over-face treatment reveals a pattern suggestive of learning. In Experiments 5–6, the subjects performed randomly on the probe trials. However, by Experiments 7–9, the subjects occasionally performed slightly or significantly above chance. Finally, in Experiment 13, the subjects performed significantly above chance on these probe trials (and, indeed, most of the other probe trials as well). Note that this general learning pattern from Experiment 1 to Experiment 13 held true for all conditions in which the rule *Gesture in front of the person whose face is visible* could work. In contrast, the subjects continued to perform randomly in the initial sessions of new conditions that instantiated seeing/not seeing by having the faces of both experimenters visible (distracted-versus-attending and eyes-open-versus-closed) (Experiments 12 and 13). An additional piece of intriguing evidence in support of this interpretation is that the subjects also chose randomly in the blindfolds treatment in Experiment 13. Unlike the other main treatments, equal amounts of both experimenters' faces were visible; the only difference was that in one case the eyes were obscured and in the other case the mouth was obscured. This, coupled with the subjects' initially random performance in the distracted-versus-attending condition (both faces visible) in Experiment 12 as well as their random performance on eyes-open-versus-closed in Experiment 13, strongly indicates that our young chimpanzees learned to use the presence of an unobscured face as a stimulus to determine the correct response.

4. The fact that the subjects tracked the experimenters' gazes in Experiment 12, but nonetheless simultaneously gestured equally often to the attending-versus-distracted experimenters, demonstrates how a phenomenon observed in natural settings (which was easily brought under experimental control in our lab) may be unrelated to an understanding of the attention that underlies the behavioral cue (the experimenter looking up) that triggers it. (In addition, this finding rules out the potential methodological criticism of our work that the chimpanzees were simply not attending to the face/eyes of the experimenters.)

5. The results of our comparative research with young children cast doubt on the possibility that the task was measuring an understanding of visual perception as a knowledge acquisition device. Young 3-year-olds, who typically perform quite poorly on tasks that require this kind of appreciation of seeing, immediately gestured selectively to the person who could see them at levels well exceeding what would be expected by chance. Even the older 2-year-olds performed above chance in most of the treatments (although see the "Discussion" section of Experiment 15 and below). This

is especially interesting because, in attempting to assess the "ecological validity" of the task, it is curious to note that a central component of the task was more relevant for the chimpanzees (e.g., begging for food) than it was for the children. The children were required to use a rather odd behavior to obtain stickers—pressing their hand (palm down) on a handprint in front of an experimenter—whereas the begging gesture that the chimpanzees used was quite natural in this context. Thus, although the task may have been more ecologically relevant for our young chimpanzees than it was for the preschool children, this did not preclude the children from immediately demonstrating an underlying mentalistic appreciation of visual perception, nor did it unveil such comprehension on the part of our young apes.

INTERPRETING THE TASK

Ultimately, a cognitive task is only as good as its ability to measure what we think it is measuring. Cole and Means (1981) have argued that one of the primary challenges facing researchers interested in experimental investigations of cognition is to develop appropriate theories of the tasks posed to their subjects. Thus, although we have reached the general conclusions outlined above, there are several aspects of our task that warrant further consideration—considerations that will force us to temper the premature conclusion that our subjects did not understand anything at all about seeing as attention. If we are to provide the best possible interpretation of the chimpanzees' performances as well as make progress toward designing future experiments that could further test the generality of our conclusions, it will be necessary carefully to examine several possible methodological reasons why our young chimpanzees might have displayed little evidence for understanding the subjective aspects of visual perception, even if they really did understand this fact about visual perception. In what follows, we do not aim to provide an exhaustive theory of our task. Rather, we seek to examine aspects of the task that may have interfered with its ability to provide a universal diagnostic of the attribution of seeing as attention.

Nature of the Subjects' Response

We begin by questioning the logic of our decision to use the chimpanzee's natural begging gesture as the behavioral response that served as our critical dependent measure. Our rationale for using the gesture was straightforward. The gesture is a naturally occurring one that these subjects use frequently in nontesting situations. In addition, it was a gesture that in the past has been used with other chimpanzees as at least functioning to designate the location of hidden objects (Povinelli, Nelson, & Boysen, 1992). Thus, we reasoned that, if young chimpanzees do possess an intentional view of others, then using one of their naturally occurring behaviors would maximize our chances of being able to demonstrate the phenomenon.

However, there are several possible limitations to this approach. First, although chimpanzees can and do learn to use their begging gestures as a means to gain access to desirable objects or people, there is no independent confirmation that they themselves comprehend this fact. Indeed, the gesture in question is used in several different contexts in natural chimpanzee societies, which rarely (if ever) include specific pointing to designate objects or organisms (Goodall, 1986; Menzel, 1974; Povinelli & Davis, 1994; Premack, 1984; but see de Waal, 1982). Thus, the mere fact that it is a "natural" gesture (which on the surface looks somewhat like pointing) does not necessarily increase the ecological validity of the task. Although using a species-typical gesture can increase the rate at which subjects learn a particular operant response (see Shettleworth, 1975), we can think of no reason why it should increase the likelihood of their comprehending the intentional significance of the gesture (assuming that such a capacity were present). For example, if chimpanzees do not really "point" proto-declaratively, then training them to use a begging gesture as an operant response to initiate an experimenter bringing them food may be comparable to training them to execute a different natural behavioral pattern (e.g., touching their elbow) to do the same. This would be the case even if the chimpanzees possessed an understanding of their gesture as *I want some of what I see.* Conversely, note that, even though the young children used what was for them an arbitrary response (pressing their hand against a piece of paper), this did not preclude them from behaving in the manner predicted by the mentalistic framework.

Young Chimpanzees as Objects of Visual Perception

A second methodological objection to our task is that in our paradigm the chimpanzees were always being asked to reason about themselves as objects of the visual perception of others. Even though the experimenters did not look directly at the chimpanzees as they entered the testing unit, our task required them to understand that, by approaching the partition and responding in the location where the experimenter's line of sight was oriented, they could become the object of the experimenter's attention. In most traditional level 1 visual perspective-taking tasks, the subjects are asked to reason about an observer looking at an inanimate object in the world (see the references in Chap. II; for exceptions, see Lempers et al., 1977, tasks 5, 13, 14). However, our decision to confront the chimpanzees with situations in which they themselves were potential objects of perception was not accidental. We reasoned that, if most organisms are sensitive to a pair of eyes looking in their general vicinity, and if our animals appreciated the subjective dimension of seeing, then they might most readily demonstrate it in a situation in which they could be the objects of such perception. In addition, given the cascading series of complex social events that spill out of direct and indirect eye contact in chimpanzee society (and of highly social nonhuman primate species in general), knowing that someone is looking at you might be one of the most primitive arenas in which we could expect to find a mental appreciation of seeing if it existed. Notice, however, that the mere fact that attention to eye gaze and eye contact exists and plays a causal role in social interactions does not by itself guarantee that such an understanding is present. We merely took as a starting point the social and evolutionary significance of being the object of someone else's visual perception.

However, despite our rationale for constructing our tests in such a manner, our chimpanzees might have had difficulty integrating that they themselves were the potential target of someone else's visual attention with the behavioral response necessary to elicit a reward. From this perspective, it could be argued that subjects of our tests must minimally understand two distinct propositions: (1) *One of the two persons can see me and is therefore subjectively connected to me,* and (2) *I should gesture toward the one who can see/is subjectively connected to me.* Perhaps the chimpanzees construct the first proposition but not the second. In this case, the chimpanzees could rightfully be said to have a mentalistic appreciation of seeing but still not demonstrate it by gesturing toward the one who can see.

Our assumption (in part supported by our results with the young children) is that this is not the case. We view statement 2 above as simply a feeling of necessity that emerges from statement 1. In other words, although these two propositions could be viewed as distinct, we assume that, if an organism appreciates that seeing is intentional, this should (by itself) subjectively connect the organism as the object of visual perception to the one who is doing the perceiving and further link their subsequent actions for that reason alone. If this assumption is true, it would not be necessary for statement 2 to exist as a formal proposition in order for subjects to gesture toward the one who can see them. The subjects would need no additional justification in order to gesture toward the one who sees them. However, we emphasize that at present this remains speculation and that further empirical work is needed.

Another way of looking at this question would be to ask whether we might expect different results if the young chimpanzees had been asked to reason about an experimenter looking at someone or something else. Would they understand that the experimenter was subjectively connected to that object or person and not another one? Some developmental psychologists have approached this problem recently with human infants. Their results can be interpreted to mean that, by 12–18 months, infants use attentional cues to locate the object of an adult's attentional focus (for a review, see Baldwin & Moses, 1994). However, the issue is a difficult one because tracking line of sight may be a fairly automatic process, dissociated from a subjective appreciation of attentional focus. Much (but not all) of the work on level 1 perspective taking with 3-year-olds by Flavell and his colleagues has centered on children's ability to use gaze direction as a means of correctly inferring what someone can or cannot see (e.g., Flavell et al., 1981; Lempers et al., 1977). Baron-Cohen (1994), however, has argued that some of these tasks can be solved by geometric calculations that do not necessarily require a subjective appreciation of attention.

To clarify the problem, let us consider a hypothetical example with chimpanzees. If, after setting the appropriate context, chimpanzees were to see an experimenter looking at one of two closed boxes and then subsequently used this gaze as a guide to discover which box contained a banana, we might still not have compelling evidence for a subjective appreciation of seeing. Algorithms of the type proposed to explain the subjects' automatic tracking of line of regard in Experiments 12 and 13 might likewise explain the subjects' responses in this circumstance. We see this as a very real problem and are currently exploring tasks that might be able to address these difficulties. Given the unexpectedly fragile performance of the youngest children in Experiment 15, such investigations are even more critical.

Finally, with respect to the question of whether young chimpanzees are able to understand themselves as objects of visual perception, it is important to ask whether our results might have been different had we chosen to have the experimenters stare directly into the chimpanzees' eyes. Would the salience of mutual gaze have increased the likelihood of the subjects' understanding that the experimenter was attentionally connected to them? Perhaps, but, as we outlined in Chapter III (see Brothers & Ring, 1992), at present we have no way of teasing apart the emotional valence of such mutual gaze (in which we were less interested) from its cognitive interpretation (in which we were most interested). Indeed, we have conducted one experiment with these same subjects in which they were allowed to choose between two experimenters, one who made direct eye contact and one who did not. Although the subjects displayed at least a weak preference to gesture in front of the one making direct eye contact, we do not conclude from this finding that these chimpanzees possess a cognitive interpretation of seeing as the mental state of attention.

Interspecific Nature of the Tasks

A third methodological issue that concerns us is that, in all our tests with the chimpanzees, the subjects were required to reason about the visual perception of members of another species. Curiously, this feature has been present in almost every experimental investigation of theory of mind abilities in nonhuman primates (Povinelli, Nelson, & Boysen, 1990, 1992; Povinelli, Parks, & Novak, 1991, 1992; Povinelli et al., 1994; Premack & Woodruff, 1978b; Woodruff & Premack, 1979; but see Cheney & Seyfarth, 1990a). Eddy, Gallup, and Povinelli (1993) noted this fact in the context of some experimental research on the influence of phylogeny on adult humans' anthropomorphic attributions. Their data indicate that humans do not attribute higher cognitive functions to animals in a random fashion; rather, their anthropomorphisms are based on their perceptions of how similar the animals are to themselves. This perceived similarity apparently has at least two strong inputs: the degree of physical similarity between themselves and the species in question and the degree to which they have formed attachment bonds with particular species. For these reasons, primates, on the one hand, and dogs and cats, on the other, were consistently judged as most likely to possess high degrees of cognitive functioning (Eddy et al., 1993).

This human attributional gradient raises direct questions about nonhumans as well. One criticism of our work (indeed, all work thus far in this area) is that the chimpanzees might display far more compelling evidence of theory of mind if they were required to reason about mental states in other chimpanzees (Cheney & Seyfarth, 1992). Translated into the terms of Eddy et al. (1993), perhaps the chimpanzees' attributional gradient is characterized by a steeper drop-off function outside members of their own species than is the case for humans. Describing the chimpanzees' attributional gradient will, in effect, result in a scientific theory of *Pan*morphism, not *anthropo*morphism. By definition, if chimpanzees do possess some aspects of theory of mind, then some of our anthropomorphisms about them must be at least partially correct because some of the subjective aspects of our psychology will turn out to be primitive traits that humans and chimpanzees inherited from a common ancestor. Thus, determining what chimpanzees (*Pan troglodytes*) know about the mind can be viewed as an attempt to determine what it would be like to possess a *Pan*morphic (as opposed to an *anthropo*morphic) view of other organisms. Owing to recent common ancestry, there may well be some overlap in these zoomorphisms, but they may turn out to be quite distinct. The general point is that it is possible to maintain that chimpanzees might perform according to the mentalistic framework when observing chimpanzee actors but not while observing human actors.

Although ultimately it is an empirical issue, we doubt the current form of the account offered above for two reasons. First, even if the general notion is correct, the relationship between captive chimpanzees and humans meets both of the criteria that appear necessary for inclusion within a hypothesized *Pan*morphism gradient: (1) captive chimpanzees have formed strong attachment bonds with humans, and (2) humans resemble them physically. Our second reason for doubting such an account is that, from a developmental perspective, theory of mind attribution by young children appears to be fairly hardwired to apply to most animate and inanimate entities, which may be suppressed only later in development (Miller & Aloise, 1989). This means that, if we wish to maintain that humans and chimpanzees both possess a common primitive developmental program controlling the intentional interpretation of behavior, then we must also argue that humans apply this interpretation broadly across the animate and inanimate universe but that chimpanzees apply it only to members of their own species. Evolutionary scenarios allowing for this state of affairs could be constructed, but we see no unique reason to favor this view. Nonetheless, although an empirical test of the general idea that our chimpanzees performed poorly because of the interspecific nature of the tasks would be difficult, it is interesting and warrants consideration.

Finally, attributing our chimpanzees' performance to the interspecific nature of the task cuts both ways. If chimpanzees' attributions are largely restricted to their own species, then why do humans differ so dramatically? That is, if humans and chimpanzees separately inherited the same developmental pathway controlling the expression of the attribution of seeing from their common ancestor, then why do humans apply it to chimpanzees but chimpanzees not apply it to humans? In the context of our experiments, perhaps the most striking illustration of this issue occurred after nearly every unsuccessful probe trial in which the subjects gestured toward the experimenter who could not see them. As their trainer lifted the screen to usher them out of the test room, he invariably said aloud, "No, Bu, he/she can't see you." This comment was completely spontaneous, and not only did he say this after the first few trials, but he continued to do so throughout the 6½ months during which the subjects were tested in this research program.

We interpret this effect as an example of the striking degree to which humans automatically process mentalistic information about seeing. We focus in particular on the trainer's unchoreographed comments because he was not directly involved in the theoretical discussions of the importance of the subjects' choices for interpretations of their theory of mind. However, all participants in this research (naive and sophisticated experimenters alike) quite frequently reported a strong surprise each time the subjects gestured toward someone who clearly could not see them. The point that we wish to make is simple. If the chimpanzees' performance was negatively affected by the interspecific nature of the experiments, the humans who participated in the experiments were not similarly affected. Their anthropomorphic gradient clearly extended to cover these young chimpanzees.

These admittedly post hoc observations about the experimenters' reactions to the chimpanzees' behavior are of interest from a slightly different perspective as well. It might be argued that the act of participating in this task practically every day for over 6 months affected the subjects' likelihood of engaging in appropriate attributions about seeing. Of course, our within- and between-session control trials were included in all experiments to rule out a general argument of this kind. However, it could be maintained that the routinization of the task somehow selectively affected the subjects' theory of mind dispositions, without affecting their performance on adjacent trials that did not require mental state attributions. However, because the subjects' trainer was present during every test session of every animal, he participated in six to seven times as many trials as any one of the individual chimpanzees; in general, the other experimenters participated in at least as many trials as any one of the chimpanzees. Yet despite this routine, which lasted nearly seven months, the trainer's spontaneous comments, as well as the experimenters' feelings of disbelief about what the subjects were doing, did not begin to dissipate until the subjects began to perform correctly in the later experiments. To us, this suggests that, although theory of mind attributions may be largely routinized and automatic, the human intentional stance (at least insofar as visual perception is concerned) is so pervasive that the detection of behavior that obviously flies in the face of such a framework does not lend itself to rapid habituation (for discussion of the "intentional stance," see Dennett, 1983).

IMPLICATIONS FOR HUMAN THEORY OF MIND DEVELOPMENT

Certain aspects of our results may raise broader questions about the necessity of appealing to an understanding of mental states to explain the behavior of infants and very young children. Indeed, our demonstration that young chimpanzees can learn (if given ample opportunity) to behave in a fashion that "mimics" how an organism with a theory of mind might behave raises questions about the extent to which the young children's performance on our task might be explained in a similar fashion. In turn, this raises an even more haunting specter for psychologists interested in theory of mind development; namely, the possibility that much of young children's performance on theory of mind tasks reflects their learning of social conventions and rules that produce the appearance, but not the reality, of an understanding of the minds of others (for descriptions of alternatives to the "theory theory" interpretation of young children's understanding of the mind, see Astington & Gopnik, 1991). In short, if we can explain our young chimpanzees' successful behavior without recourse to positing an understanding of mental states, then perhaps similar accounts of the young children's behavior can likewise be posited.

Young children's use of mental state terms raises an analogous problem. For example, Bartsch and Wellman (1995) argue from an examination of natural language use that by 18–24 months young children already have a reasonable understanding of the mental state of desire. In contrast, and consistent with the view being examined here, Gordon (1996) argues that perhaps their correct use of mental state terms can be explained by mental ascent routines by which they figure out the appropriate mental states terms to use in the right contexts.

In order to explore this problem more carefully, we examine our own tests as a case in point. We cannot overlook the fact that the young children (even the youngest ones) who participated in our tests had prior experience with occlusion of eyes in the context of both general day-to-day experiences with their caregivers and peers and more specific experience in the context of games. Thus, a skeptic of the extent of the young children's understanding of the mental dimension of our tasks could claim that the children had already received the necessary "learning trials" long before they participated in our tests (albeit not in the same structured fashion), thus accounting for their successful Trial 1 performance. To some extent, we have already addressed this issue earlier by emphasizing the arbitrary nature of the task for the young children. However, it is equally important not to overlook the fact that our chimpanzees had these two types of experiences with visual occlusion as well, in interactions with both humans and other chimpanzees. Although we are not trying to claim that they had exactly the same experiences as young children, we would maintain that their experiences were sufficiently similar to allow them to discover many of the consequences of visual occlusion. For instance, they (like many other primates) engage in simple "deceptive" acts in which they use their bodies to occlude objects from others, they spend hours playing chase games into and out of areas that result in visual occlusion, and (most critically) they also use objects or their hands or arms to produce visual occlusion in themselves and then explore the consequences by moving slowly across the compound until they collide with a solid object (see Chap. III).

Given this rich backdrop of experiences with visual occlusion on the part of both the children and the chimpanzees, we are forced to ask whether tasks such as the ones that we administered are sensitive enough to tease apart what the subjects (children and apes alike) have learned through prior experiences from what they genuinely conceptualize about vision as an attentional process. More generally, for any behavior that might emerge as

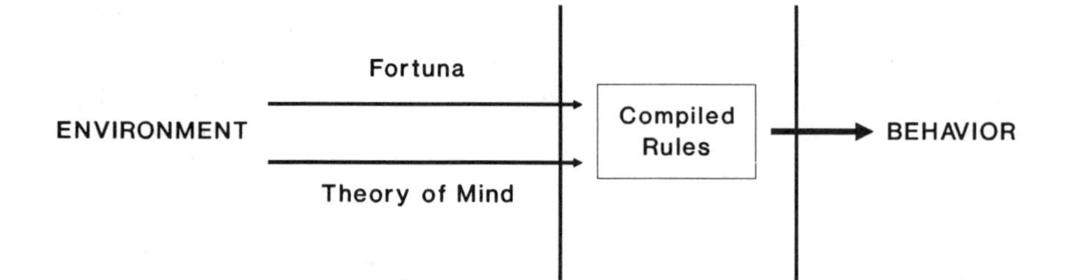

FIG. 28.—Alternative processes for generating complex social behavior. Most organisms have tightly canalized epigenetic pathways that generate algorithms for social behavior. In addition, social behaviors can also be generated through differential feedback from other group members (learning through trial and error). Finally, humans (at least) use internally generated representations of the mental states of others to produce rapid behavioral solutions to social problems. In principle, the exact same behavioral patterns could emerge from any of these pathways, making the determination of which has been used difficult to discern. In the absence of a complete knowledge of an organism's developmental history and a complete understanding of its generalization abilities, distinguishing between "Fortuna" and "theory of mind" processes poses great challenges.

the result of an inference about a mental state, the exact same behavior may emerge as the result of learning rules that do not appeal to mental states. A third possibility also exists. Organisms may first possess conceptual knowledge of the specific mental states but must learn how to apply them correctly (Povinelli, 1993). Further complicating the picture is that situations that organisms originally decoded within a theory of mind framework may through time be compiled into a series of more or less automatically executed rules. Obviously, learned rules may be compiled in a similar fashion. These ideas are represented in Figure 28. Cast in these terms, this would seem to be a methodological question directed at our current techniques, not a theoretical question challenging the notion of theory of mind in young children or chimpanzees.

While recognizing the overall viability of this line of thinking, we advocate a moderate reaction to this problem. First and foremost, although at the outset of these studies we dichotomized the possible explanations of chimpanzees' and children's performance as being based in either learning or theory of mind, this dichotomy is clearly only a first heuristic approximation. Indeed, this dichotomy is probably useful only in the atmosphere of our currently crude understanding of how theory of mind concepts become implanted in the developing mind. Clearly, the acquisition and elaboration of mental state concepts depends on the ability to learn relations among animate and inanimate objects and how they can causally interact to produce changes in the world. Indeed, the real utility of mental state concepts may be that they dramatically reduce the search space required to solve social problems. Concepts such as attention, desire, knowledge, and belief all provide readily available folk explanations for why organisms have behaved in a certain way or reasons for predicting how they will behave in the future. Thus, instead of having to search through an entire field of assembled objects, organisms, and events, organisms with a theory of mind have a powerful cognitive framework for extracting the relevant stimuli in the causal chain of events of interest. From an evolutionary perspective, these mental state concepts can easily be understood as being advantageous in a social context because they allow for more accurate descriptions of reality by the organism that deploys them. After all, organisms are controlled by the representations of reality stored in their neural circuits, even if they are completely unaware of this fact. To construct concepts of such mental states hardly requires the evolution of an understanding of how the brain supports such representations; most of the time, a folk psychology of mental state concepts will serve the organism quite well. In this sense, learning processes and theory of mind attributions are not dichotomous, although there may well still be a qualitative distinction between those organisms that possess such mental state concepts and those that do not.

Viewed in this light, there are several reasons to defend our techniques as sensitive enough to distinguish between the two processes in Figure 28. First, we intentionally designed most of our tasks so that they would not trigger the deployment of compiled rules (the one notable exception was the test of gaze following in Experiment 12). Instead, we created circumstances where an organism could recombine existing behaviors into a correct solution after surveying a novel problem: choosing to beg from only one of two experimenters. We believe that a fair evaluation of the structure of the tasks reveals that the problem was as novel (or arbitrary) for the apes as it was for the children (see above). Given the challenge of distinguishing between learned rules that take mental states as their content and those that do not, the best diagnostic is how the subjects respond on the first couple of trials. Indeed, the best evidence would emerge from an assessment of Trial 1 performance. In this regard, the children's success at generalizing to this novel task on Trial 1, juxtaposed against the failure of the chimpanzees to do likewise, in and of itself suggests that the children were bringing a different interpretive framework (a theory of mind) to bear on the problems that we posed to them. When viewed in this light, our results showing that chimpanzees might learn to behave in ways that mimic possessing a theory of mind are not surprising at all, nor do they provide a theoretical challenge to theory of mind research in general. On the other hand, other aspects of the results can and do raise questions about whether specific behaviors (such as gaze following) should be considered diagnostic of an understanding of attention in others in either human infants or chimpanzees.

There is a final issue that is more difficult to resolve. If by participating in our experiments our chimpanzee subjects learned (or were learning) relations about eye gaze and contact so well that they could now generalize it within one to four trials to most new situations that we created for them, we are left with a troubling thought. There may be no way to distinguish between three possibilities: (1) the subjects did not originally appreciate the existence of the mental state of attention, but now they do, and they understand one of its appropriate sensory bases; (2) the subjects originally possessed an amodal theory of attention but, again, now have localized one of its appropriate sensory bases; and, finally, (3) the subjects neither began nor emerged from our tests with a subjective theory of attention; rather, as a consequence of our testing (read: training) they were able to perform *as if* they possessed such a theory. At present, we have no clear proscription for a research strategy that could disprove any of these alternatives. To be sure, new subjects could be used, but after similar testing they would be in a similar position. Future methodological and theoretical inquiries must continue to grapple with this problem.

WHAT CHIMPANZEES MIGHT KNOW ABOUT SEEING AND ATTENTION: ALTERNATIVE HYPOTHESES

If our results do not lend themselves to a mentalistic interpretation of young chimpanzees' understanding of visual perception, it is necessary to consider the various alternative hypotheses that might be consistent with the data that we collected. In doing so, we offer some ideas about the kinds of information that chimpanzees must process about the eyes and what kinds of behavior such processing supports. Although it is impossible at this point to specify the exact variables that control eye monitoring on the part of chimpanzees, the phenomenon is widespread in the social lives of chimpanzees as well as other social mammals (e.g., Goodall, 1968). Thus, each of the alternative hypotheses that we consider next must allow room for such monitoring and use of eye gaze while simultaneously accounting for our data. We can think of at least three plausible hypotheses that fit these conditions.

Hypothesis 1: Older Chimpanzees Develop a Subjective Understanding of Seeing

First, it is possible that chimpanzees older than 6 years of age develop a folk theory of visual perception. Simply because our 5–6-year-old chimpanzees displayed no evidence that they understood the mental significance of seeing, this does not eliminate the possibility that older members of the species might. By analogy, if we were to test 9-month-old human children on tasks that required them to take into account the false beliefs of others, we could not rightfully use their absence of comprehension to claim that the human species does not understand the representational property of belief. Indeed, studies of comparative cognitive development provide data that make the possibility that older chimpanzees may display evidence of understanding the subjective aspects of seeing at least plausible. There is mounting evidence that, by the end of the first year of life, chimpanzee psychological development is markedly delayed relative to humans (or, conversely, that human development is accelerated relative to chimpanzees). Independent researchers have obtained data that sensorimotor intelligence, vocal comprehension skills, mirror self-recognition, and spontaneous second-order classification skills may not emerge in chimpanzees until 5–8 years of age (Chevalier-Skolnikoff, 1983; Mathieu & Bergeron, 1981; Mignault, 1985; Povinelli et al., 1993; Savage-Rumbaugh et al., 1993; Spinozzi, 1993). Yet all these abilities that seem to emerge at around 4–6 years or so in chimpanzees typically emerge at about 18–24 months of age in human children.

Thus, if a genuinely mentalistic appreciation of seeing does not emerge until 2–2½ years of age in human children, and if chimpanzee psychological development is slower than human psychological development, then we might not expect to obtain evidence for this ability until chimpanzees are somewhat older than 6 years of age. Of course, this reasoning hinges on several key assumptions. First, it assumes that the last common ancestor of chimpanzees and humans possessed a developmental program that included the construction of a mentalistic appreciation of seeing (and, more likely, an intentional interpretation of behavior generally). It also assumes that, during the separate evolution of the chimpanzee and the human lineage, that ancestral program was either temporally compressed in humans or extended in chimpanzees (for discussions of the likely patterns of heterochrony in the physical evolution of the great apes and humans, see McKinney & McNamara, 1991; Shea, 1983, 1988). A final (and perhaps most controversial) assumption is that there is little interspecific décalage among these skills. In other words, it assumes that there are necessary reasons why, for example, a closure of sensorimotor development must precede the development of an intentional understanding of others.

If the framework outlined above has some merit, then we ought to be able to use various (at this point, arbitrary) markers of cognitive development across the two species in order to determine whether such an account makes sense for our subjects. Unfortunately, we do not have a solid database for our subjects across many traditional measures of cognitive development known to overlap in the two species. However, we do have longitudinal data on one such measure: self-recognition in mirrors. Five of the subjects in the experiments reported in the earlier chapters were tested for self-recognition at three distinct time points: approximately 3, 3½, and 4½ years of age (for details of these studies, see Povinelli et al., 1993). The other two subjects have fewer time points (Jadine, 3 and 4½ years; Candy, 3 years).

These longitudinal investigations demonstrated that only one subject (Megan) showed evidence of self-recognition from 3 years of age forward. Two other subjects (Jadine, Mindy) showed arguable evidence of self-recognition by 4½ years of age. Three of the other subjects (Brandy, Kara, Apollo) have shown no compelling evidence to date of self-recognition, including during some rather dramatic unpublished tests in which their hair was surreptitiously dyed a bright orange color. The final subject (Candy) was only formally tested once at about 3 years and was diagnosed as negative. The general point that can be extracted from these data is that our subjects displayed little evidence for appreciating the mentalistic significance of seeing and also have displayed the characteristic pattern of slower psychological development that is arguably true of chimpanzees in general. Thus, if self-recognition emerges as a consequence of psychological developments that occur at around 18 months in human infants, and if a genuinely mentalistic appreciation of seeing must await those developments, then the late onset of self-recognition in our subjects can be interpreted as a reason to hold out for a still later emergence of a mentalistic theory of seeing (and behavior in general) in the species.

We caution, however, that we are not overly optimistic on this point precisely because we failed to detect a relation between our chimpanzees' self-recognition status and their performance on a task that would seem to be measuring an ability that (at least in humans) is one of the earliest expressions of a knowledge about the subjective experiences of others. In contrast, in humans there is some reason to believe that the onset of mirror self-recognition may be correlated with the development of a number of other capacities related to theory of mind (Asendorpf & Baudonniere, 1993; Bischof-Köhler, 1988; Johnson, 1982; Lewis, Sullivan, Stanger, & Weiss, 1989). From one perspective, such a correlation is exactly what was predicted by Gallup (1982; Gallup & Suarez, 1986) a number of years ago. However, if many of these emerging skills or abilities are yoked to broader developmental constructions, then the extent to which an underlying form of self-conception deterministically expresses its presence both as self-recognition in mirrors and in behaviors suggestive of theory of mind (Gallup's hypothesis) becomes less clear.

One alternative to Gallup's (1982) view is that, in humans at least, primitive theory of mind abilities and either self-conception or some other abilities supporting mirror self-recognition emerge at roughly the same age but are not causally related. One way of testing Gallup's model against this alternative is to test other species that display self-recognition for the earliest emerging theory of mind abilities—abilities that may be found in studies of their understanding of what Flavell refers to as level 1 visual perspective taking (Povinelli, 1993; Povinelli & deBlois, 1992; Povinelli et al., 1991). The research reported in this *Monograph* can be seen as exactly this kind of test of Gallup's model. However, our young chimpanzees—some of whom had shown evidence of mirror self-recognition for up to 2 years prior to our current tests, fared no better than their companions. The results, therefore, do not provide any support for this revision of Gallup's model. Indeed, they are evidence against the model. Thus, although it remains possible that older chimpanzees may develop a truly mentalistic appreciation of seeing, we are not overly optimistic on this point.

Hypothesis 2: Chimpanzees Possess an Amodal Theory of Attention

A very different hypothesis about young chimpanzees' theory of mind is that they do possess an appreciation of how subjective experiences of others ground them to the world but that their theory does not appeal to particular sensory modalities as the basis of such a connection. In other words, chimpanzees may possess an amodal theory of subjective attention. Perhaps for them, other organisms possess subjective states such as attention, but these states have no particular sensory basis. By analogy, consider Lyon's (1993) argument that children's earliest understanding of knowledge in others is based on their interests or desires, not their perceptual contact with information. In an elegant pair of studies, he was able to demonstrate that young 3-year-olds attributed knowledge to others not on the basis of whether they had proper perceptual contact with a situation but rather on the basis of whether they had displayed prior interest in the situation.

Our argument about young chimpanzees' understanding of the mental state of attention can be cast in similar terms. For chimpanzees, perceptual contact through an individual sensory modality per se may not be needed to attribute a subjective connection to the world via attention. In the context of the present research, our efforts to have both experimenters configured in exactly the same fashion (except for differences that were dictated by creating the various examples of "seeing" and "not seeing") may have created a situation in which both experimenters appeared equally interested in what was occurring, despite the fact that one could not see. This would, in effect, deny an organism with an amodal theory of attention the appropriate behavioral cues that are needed to attribute attention.

Such an interpretation casts a slightly different light on the subjects' differential performance in the back-versus-front treatment as compared to the others. Perhaps instead of (or in addition to) the stimulus similarity hypothesis that we offered to explain the subjects' success in back-versus-front, this condition simply provides the subjects with obvious cues of interest-versus-disinterest. In contrast, in the other conditions, the subjects may have attributed equal interest to both experimenters because they were both facing forward in similar postures. If chimpanzees' attribution of attention is not based on the access of specific modalities to a given situation, then, for example, the selective deprivation of one experimenter's eyes as opposed to the other's ears (hands-over-eyes treatment) may have no effect on the subject's willingness to attribute a subjective connection to the situation. Likewise, Lyon's (1993) 3-year-old children were quite likely to attribute knowledge about a toy's identity to someone who could not see it (because he or she was blindfolded) but showed a desire to know about it, as opposed to someone who could see the toy but indicated that he or she did not want it. This emphasis on desire and interest in a situation over perceptual access in forming knowledge attributions does not make the attributions any less "subjective" or "mental," although they may not qualify as attributions of justified, true belief.

If this hypothesis is correct, then it must be wedded to a learning interpretation to explain the results of our later experiments. After all, although the subjects performed randomly on most treatments early on (which might be predicted by the hypothesis under consideration here), they performed much better in the final investigations. Thus, it would still be necessary to argue that the subjects learned something across the experiments. One possibility is what we outlined earlier: the subjects learned a rule about gesturing in front of the person whose face (or possibly eyes) was visible. Another possibility is that they learned (or at least were learning) something about their own attributions: namely, that the only experimenter subjectively connected to them was the one who could see. If this were the case, the subjects could be thought of as starting with an amodal theory of attention and gradually narrowing in on "seeing" as a key factor for determining attentional focus. As mentioned in Experiment 13, Megan is a likely candidate for this position.

Hypothesis 3: Theory of Mind Is a Uniquely Derived Feature of the Human Species

The final hypothesis with which our data are consistent is that a mentalistic understanding of visual perception in particular, or an intentional understanding of others (a theory of mind) more generally, is an autapomorphic trait of the human lineage. Simply put, theory of mind represents a unique innovation that occurred during the course of human evolution. This view is diagrammatically represented in the "late evolution" model presented in Figure 2 above. If this hypothesis is correct, then previous findings that have been interpreted as consistent with the view that chimpanzees do engage in mental state attribution would have to be reinterpreted in other ways.

As we have seen in Chapter II, several such previous investigations could be interpreted in this way (see the discussions in Povinelli, 1991, 1993, 1996, and Heyes, 1993, concerning, e.g., the findings reported by Povinelli et al., 1990, and Premack, 1988). The data reported in Chapter III and IV of this *Monograph* illustrate the potential heuristic value of such an account of previous results. On the heels of criticisms by Premack (personal communication, 20 February 1990), we have presented our subjects' performances in blocks of two trials in most graphs in this *Monograph,* and here (and elsewhere) the results are able clearly to distinguish between conditions that the chimpanzees immediately understood and those that they did not (see also Povinelli, Nelson, & Boysen, 1992, p. 637). As discussed above, a meta-analysis across all experiments in this *Monograph* reveals that, in a number of cases, learning (in the context of the tests that we administered) could account for the changes in the subjects' responses to some, but not all, of the treatments (e.g., back-versus-front). The operation of such learning processes in the natural world of monkeys and apes and other mammals (where, after all, they first evolved), in combination with the development of tightly canalized social behaviors and dispositions, could provide a unifying account of the rich anecdotal database concerning primate social manipulation and deception (e.g., Whiten & Byrne, 1988).

In this context, what are we to make of Premack and Woodruff's (1978b) quip that it would "waste the behaviorist's time to recommend parsimony to the ape. The ape could only be a mentalist. Unless we are badly mistaken, he is not intelligent enough to be a behaviorist" (p. 526)? Our reply, of course, is an empirical one. If, under the advancing scrutiny of improved experimental methods, the chimpanzee's theory of mind turns out to be a behavioral illusion based (in part) on our subconscious detection of ancient physical and behavioral traits in chimpanzees that we still possess, then we will be forced to conclude—without undue prejudice—that they simply do not share with us a suite of psychological dispositions that evolved exclusively in the human lineage. If all our mental state anthropomorphisms about chimpanzees collapse under careful investigation and we are forced to accept this third hypothesis, then it will be unfounded to think that chimpanzees *Pan*morphize us, or anyone else, at all. That is, Premack and Woodruff may have been correct that chimpanzees are not intelligent enough to be behaviorists but wrong in that they may not be intelligent enough to be mentalists either. In proper evolutionary terminology, chimpanzees may not be closely related enough to humans to have theoretical commitments one way or the other.

CONCLUSION

Our final thoughts concerning these studies are methodological in nature. The results demonstrate to us that appropriate experimental designs, coupled with sufficiently large sample sizes, can provide a very sensitive analysis of what nonhuman primates know about the mind. The young chimpanzees in these studies might have performed very differently than they did. Instead of consistently performing according to the predictions of a model that assumed that they had no theory of mind, they could have just as easily—like most of the preschoolers we tested—performed according to a different model that assumed that they had some knowledge about the relation of the eyes to an internal mental world. However, it is important to keep in mind the set of alternative hypotheses outlined above. Some combination of new procedures and/or older animals may yield different results. Nonetheless, until additional data emerge to supplant those reported here, we conclude that, despite their striking use of (and interest in) the eyes, 5–6-year-old chimpanzees apparently see very little behind them.

APPENDIX

TABLE A1

List and Description of Baseline and Treatment Probe Trials Used in Experiments 1–14 for Testing Young Chimpanzees' Understanding of Seeing

Probe Trial Type	Description of Stimuli	Experiments
Baseline		
Baseline (A)	One experimenter offering block of wood, other offering food reward	1–6 (standing) 7–12 (sitting)
Baseline-plus-attending-versus-distracted (A′)	Same as A, except experimenter offering food looks above and behind subject	12 (sitting)
Probes using objects to cause visual occlusion		
Blindfolds (B)	One experimenter has blindfolds over eyes, other has blindfolds over mouth	1–2 (standing) 13 (sitting)
Buckets (B′)	One experimenter holds bucket over head, other holds bucket on shoulder	1 (standing) 13 (sitting)
Screen-over-face (B″)	One person holds screen above shoulder, other covers face with screen	5–6 (standing) 7–9, 13 (sitting)
Face-without-eyes-versus-eyes-only (D)	One experimenter with eyes closed, other looks through eyeholes cut through screen	14 (sitting)
Eyes-without-face-versus-no eyes/no face (D′)	Both experimenters hold screens with eyeholes over face; one with eyes open, other with eyes closed	14 (sitting)
Face-without-eyes-versus-no eyes/no face (D″)	Both experimenters have blindfolds over eyes; one obscures face with screen with eyeholes, other does not	14 (sitting)

TABLE A1 (*Continued*)

Probe Trial Type	Description of Stimuli	Experiments
Naturalistic treatments		
Back-versus-front (C)	One experimenter faces forward, other faces backward	1–3, 5–6 (standing) 7–13 (sitting)
Back-versus-front with screens hanging from neck (C_1)	Same as C, except both experimenters have screens (see B″) hanging from neck	5 (standing)
Back-versus-front with screens above shoulders (C_2).	Same as C, except both experimenters hold screens (see B″) above shoulders	5–6 (standing) 7–9 (sitting)
Hands-over-eyes (C′)	One experimenter covers eyes with palms, other covers ears with palms	1–3 (standing) 13 (sitting)
Looking-over-shoulder (C″) . .	Both experimenters have backs facing subject, but one looks over shoulder toward test unit, other does not	3 (standing) 10–11, 13 (sitting)
Attending-versus-distracted (C‴)	Both experimenters face subject, but one looks above and behind subject	12 (sitting)
Eyes-open-versus-closed (C⁗)	Both experimenters face subject, one with eyes open, other with eyes closed	13 (sitting)
Mixed treatments		
Back-versus-front + blindfolds (C+B)	Both experimenters have blindfolds over mouth, one faces forward, other faces backward	4 (standing)

REFERENCES

Alexander, R. D. (1974). The evolution of social behavior. *Annual Review of Systematics and Ecology,* **5,** 325–383.

Anderson, J. R. (1984). Monkeys with mirrors: Some questions for primate psychology. *International Journal of Primatology,* **5,** 81–98.

Antinucci, F. (1989). *Cognitive structures and development in nonhuman primates.* Hillsdale, NJ: Erlbaum.

Antinucci, F. (1990). The comparative study of cognitive ontogeny in four primate species. In S. T. Parker & K. R. Gibson (Eds.), *"Language" and intelligence in monkeys and apes.* New York: Cambridge University Press.

Argyle, M., & Cook, M. (1976). *Gaze and mutual gaze.* Cambridge: Cambridge University Press.

Asendorpf, J. B., & Baudonniere, P.-M. (1993). Self-awareness and other-awareness: Mirror self-recognition and synchronic imitation among unfamiliar peers. *Developmental Psychology,* **29,** 88–95.

Astington, J. W., & Gopnik, A. (1991). Theoretical explanations of children's understanding of the mind. *British Journal of Developmental Psychology,* **9,** 7–31.

Baldwin, D. A. (1991). Infants' contribution to the achievement of joint reference. *Child Development,* **63,** 875–890.

Baldwin, D. A., & Moses, L. J. (1994). Early understanding of referential intent and attentional focus: Evidence from language and emotion. In C. Lewis & P. Mitchell (Eds.), *Children's early understanding of mind.* Hillsdale, NJ: Erlbaum.

Baron-Cohen, S. (1994). How to build a baby that can read minds: Cognitive mechanisms in mindreading. *Current Psychology of Cognition,* **13,** 513–552.

Bartsch, K., & Wellman, H. M. (1995). *Children talk about the mind.* New York: Oxford University Press.

Bennett, J. (1978). Some remarks about concepts. *Behavioral and Brain Sciences,* **1,** 557–560.

Berndt, T. J., & Berndt, E. G. (1975). Children's use of motives and intentionality in person perception and moral judgment. *Child Development,* **46,** 904–912.

Bernstein, I. S. (1988). Metaphor, cognitive belief, and science. *Behavioral and Brain Sciences,* **11,** 247–248.

Bischof-Köhler, D. (1988). Uber der Zusammenhang von Empathie und der Fahigkeit, sich im Spiegel zu erkennen [On the association between empathy and ability to recognize oneself in the mirror]. *Schhweizerische Zeitschrift fur Psychologie,* **47,** 147–159.

Bitterman, M. E. (1960). Toward a comparative psychology of learning. *American Psychologist,* **15,** 702–712.

Bitterman, M. E. (1975). The comparative analysis of learning. *Science,* **188,** 699–709.

Boakes, R. A. (1984). *From Darwin to behaviourism.* Oxford: Oxford University Press.

Borke, H. (1971). Interpersonal perception of young children: Egocentrism or empathy? *Developmental Psychology,* **5,** 263-269.

Bretherton, I., & Bates, E. (1979). The emergence of intentional communication. In I. Uzgiris (Ed.), *New directions for child development* (Vol. **4**). San Francisco: Jossey-Bass.

Bretherton, I., & Beeghly, M. (1982). Talking about internal mental states: The acquisition of an explicit theory of mind. *Developmental Psychology,* **18,** 906–921.

Brothers, L., & Ring, B. (1992). A neuroethological framework for the representation of minds. *Journal of Cognitive Neuroscience,* **4,** 107–118.

Burger, J., Gochfeld, M., & Murray, B. G., Jr. (1991). Role of a predator's eye size in risk perception by basking black iguana, *Ctenosaura similis. Animal Behaviour,* **42,** 471–476.

Burghardt, G. M. (1988). Anecdotes and critical anthropomorphism. *Behavioral and Brain Sciences,* **11,** 248–249.

Burghardt, G. M., & Gittleman, J. L. (1990). Comparative behavior and phylogenetic analyses: New wine, old bottles. In M. Bekoff & D. Jamieson (Eds.), *Interpretation and explanation in the study of animal behavior: Vol. 2. Explanation, evolution, and adaptation.* Oxford: Westview.

Burghardt, G. M., & Greene, H. W. (1988). Predator simulation and duration of death feigning in neonate hognose snakes. *Animal Behaviour,* **36,** 1842–1844.

Butterworth, G., & Cochran, E. (1980). Towards a mechanism of joint visual attention in human infancy. *International Journal of Behavioral Development,* **3,** 253–272.

Butterworth, G., & Jarrett, N. (1991). What minds have in common is space: Spatial mechanisms serving joint visual attention in infancy. *British Journal of Developmental Psychology,* **9,** 55–72.

Byrne, R. W., & Whiten, A. (1985). Tactical deception of familiar individuals in baboons (*Papio ursinus*). *Animal Behavior,* **33,** 669-673.

Byrne, R. W., & Whiten, A. (1991). Computation and mindreading in primate tactical deception. In A. Whiten (Ed.), *Natural theories of mind.* Oxford: Blackwell.

Call, J., & Tomasello, M. (1994). The production and comprehension of referential pointing by orangutans (*Pongo pygmaeus*). *Journal of Comparative Psychology,* **108,** 307–317.

Campos, J. J. (1983). The importance of affective communication in social referencing. *Merrill-Palmer Quarterly,* **29,** 83–87.

Chance, M. R. A. (1967). Attention structure as the basis of primate rank orders. *Man,* **2,** 503–518.

Chandler, M. J., Fritz, A. S., & Hala, S. (1989). Small-scale deceit: Deception as a marker of two-, three-, and four-year-olds' early theories of mind. *Child Development,* **60,** 1263–1277.

Chandler, M. J., Greenspan, S., & Barenboim, C. (1973). Judgments of intentionality in response to videotaped and verbally presented moral dilemmas: The medium is the message. *Child Development,* **44,** 315–320.

Cheney, D. L., & Seyfarth, R. M. (1990a). Attending to behaviour versus attending to knowledge: Examining monkeys' attribution of mental states. *Animal Behaviour,* **40,** 742–753.

Cheney, D. L., & Seyfarth, R. M. (1990b). *How monkeys see the world.* Chicago: University of Chicago Press.

Cheney, D. L., & Seyfarth, R. M. (1992). Characterizing the mind of another species. *Behavioral and Brain Sciences,* **15,** 172–179.

Cheney, D. L., Seyfarth, R. M., & Smuts, B. (1986). Social relationships and social cognition in nonhuman primates. *Science,* **234,** 1361–1366.

Chevalier-Skolnikoff, S. (1983). Sensorimotor development in orang-utans and other primates. *Journal of Human Evolution,* **12,** 545–561.

Cole, M., & Means, B. (1981). *Comparative studies of how people think.* Cambridge, MA: Harvard University Press.
Cole, S., Hainsworth, F. R., Kamil, A. C., Mercier, T., & Wolf, L. L. (1982). Spatial learning as an adaptation in hummingbirds. *Science,* **217,** 655–657.
Darwin, C. (1982). *The descent of man.* New York: Modern Library. (Original work published 1871)
Dennett, D. C. (1978). Beliefs about beliefs. *Behavioral and Brain Sciences,* **1,** 568–570.
Dennett, D. C. (1983). Intentional systems in cognitive ethology: The "Panglossian paradigm" defended. *Behavioral and Brain Sciences,* **6,** 343–390.
de Waal, F. B. M. (1982). *Chimpanzee politics: Power and sex among apes.* New York: Harper & Row.
Eddy, T. J., Gallup, G. G., Jr., & Povinelli, D. J. (1993). Attribution of cognitive states to animals: Anthropomorphism in comparative perspective. *Journal of Social Issues,* **49,** 87–101.
Epstein, R., Lanza, R. P., & Skinner, B. F. (1981). "Self-awareness" in the pigeon. *Science,* **212,** 695-696.
Fehr, B. J., & Exline, R. B. (1987). Social visual interaction: A conceptual and literature review. In A. W. Siegman & S. Feldstein (Eds.), *Nonverbal behavior and communication* (2d ed.). Hillsdale, NJ: Erlbaum.
Feinman, S. (1982). Social referencing in infancy. *Merrill-Palmer Quarterly,* **28,** 445–470.
Feinman, S., Roberts, D., Hsieh, K., Sawyer, D., & Swanson, D. (1992). A critical review of social referencing in infancy. In S. Feinman (Ed.), *Social referencing and the social construction of reality in infancy.* New York: Plenum.
Fishbein, H. D., Lewis, S., & Keiffer, K. (1972). Children's understanding of spatial relations: Coordination of perspectives. *Developmental Psychology,* **7,** 21–33.
Flavell, J. H. (1974). The development of inferences about others. In T. Mischell (Ed.), *Understanding other persons.* Oxford: Blackwell.
Flavell, J. H. (1988). From cognitive connections to mental representations. In J. W. Astington, P. L. Harris, & D. R. Olson (Eds.), *Developing theories of mind.* Cambridge: Cambridge University Press.
Flavell, J. H., Everett, B. A., Croft, K., & Flavell, E. R. (1981). Young children's knowledge about visual perception: Further evidence for the level 1–level 2 distinction. *Developmental Psychology,* **17,** 99–103.
Flavell, J. H., Flavell, E. R., Green, F. L., & Wilcox, S. A. (1980). Young children's knowledge about visual perception: Effect of observer's distance from target on perceptual clarity of target. *Developmental Psychology,* **16,** 10–12.
Flavell, J. H., Green, F. L., & Flavell, E. R. (1989). Young children's ability to differentiate appearance-reality and level 2 perspectives in the tactile modality. *Child Development,* **60,** 201–213.
Flavell, J. H., Green, F. L., & Flavell, E. R. (1993). Children's understanding of the stream of consciousness. *Child Development,* **64,** 387–398.
Flavell, J. H., Green, F. L., & Flavell, E. R. (1995). Young children's knowledge about thinking. *Monographs of the Society for Research in Child Development,* **60**(1, Serial No. 243).
Flavell, J. H., Shipstead, S. G., & Croft, K. (1978). What young children think you see when their eyes are closed. *Cognition,* **8,** 369–387.
Frye, D., & Moore, C. (Eds.). (1991). *Children's theories of mind: Mental states and social understanding.* Hillsdale, NJ: Erlbaum.
Gallup, G. G., Jr. (1970). Chimpanzees: Self-recognition. *Science,* **167,** 86–87.
Gallup, G. G., Jr. (1975). Toward an operational definition of self-awareness. In R. H. Tuttle (Ed.), *Socio-ecology and psychology of primates.* The Hague: Mouton.

Gallup, G. G., Jr. (1982). Self-awareness and the emergence of mind in primates. *American Journal of Primatology,* **2,** 237–248.

Gallup, G. G., Jr. (1983). Toward a comparative psychology of mind. In R. E. Mellgren (Ed.), *Animal cognition and behavior.* New York: North-Holland.

Gallup, G. G., Jr. (1985). Do minds exist in species other than our own? *Neuroscience and Biobehavioral Reviews,* **9,** 631–641.

Gallup, G. G., Jr., Nash, R. F., & Ellison, A. L., Jr. (1971). Tonic immobility as a reaction to predation: Artificial eyes as a fear stimulus for chickens. *Psychonomic Science,* **23,** 79–80.

Gallup, G. G., Jr., & Suarez, S. D. (1986). Self-awareness and the emergence of mind in humans and other primates. In J. Suls & A. G. Greenwald (Eds.), *Psychological perspectives on the self* (Vol. **3**). Hillsdale, NJ: Erlbaum.

Gaulin, S. J. C. (1992). Evolution of sex differences in spatial ability. *Yearbook of Physical Anthropology,* **35,** 125–151.

Goldman-Rakic, P. S., & Preuss, T. M. (1987). Wither comparative psychology? *Behavioral and Brain Sciences,* **10,** 666–667.

Gómez, J. C. (1991a). The emergence of intentional communication as a problem-solving strategy in the gorilla. In S. T. Parker & K. R. Gibson (Eds.), *"Language" and intelligence in monkeys and apes.* New York: Cambridge University Press.

Gómez, J. C. (1991b). Visual behaviour as a window for reading the mind of others in primates. In A. Whiten (Ed.), *Natural theories of mind: Evolution, development and simulation of everyday mindreading.* Cambridge: Blackwell.

Goodall, J. (1968). Expressive movements and communication in free-ranging chimpanzees: A preliminary report. In P. Jay (Ed.), *Primates: Studies in adaptation and variation.* New York: Holt, Rinehart & Winston.

Goodall, J. (1986). *The chimpanzees of Gombe: Patterns of behavior.* Cambridge, MA: Harvard University Press.

Gopnik, A., & Graf, P. (1988). Knowing how you know: Young children's ability to identify and remember the sources of their beliefs. *Child Development,* **59,** 1366–1371.

Gopnik, A., Melzhoff, A. N., & Esterly, J. (1995, February). *Young children's understanding of visual perspective-taking.* Poster presented at the first annual Theory of Mind Conference, Eugene, OR.

Gordon, R. M. (1996). Comprehending and uncomprehending self-reference. In P. Carruthers & P. Smith (Eds.), *Theories of theories of mind.* Cambridge: Cambridge University Press.

Gutkin, D. C. (1972). The effect of systematic story changes on intentionality in children's moral judgments. *Child Development,* **43,** 187–195.

Harman, G. (1978). Studying the chimpanzee's theory of mind. *Behavioral and Brain Sciences,* **1,** 576–577.

Harris, P. L. (1989). *Children and emotion: The development of psychological understanding.* Oxford: Blackwell.

Harris, P. L. (1991). The work of the imagination. In A. Whiten (Ed.), *Natural theories of mind: Evolution, development and simulation of everyday mindreading.* Cambridge: Blackwell.

Heyes, C. M. (1988). The distant blast of Lloyd Morgan's canon. *Behavioral and Brain Sciences,* **11,** 256–257.

Heyes, C. M. (1993). Anecdotes, training, trapping and triangulating: Do animals attribute mental states? *Animal Behaviour,* **46,** 177–188.

Hodos, W., & Campbell, C. B. G. (1969). Scala naturae: Why there is no theory in comparative psychology. *Psychological Review,* **76,** 337–350.

Imamoğlu, E. O. (1975). Children's awareness in usage of intention cues. *Child Development,* **57,** 567–582.

Irwin, D. M., & Moore, S. (1971). The young child's understanding of social justice. *Developmental Psychology,* **5,** 406–410.

Itakura, S. (1987a). Mirror guided behavior in Japanese monkeys (*Macaca fuscata fuscata*). *Primates,* **28,** 149–161.

Itakura, S. (1987b). Use of a mirror to direct their responses in Japanese monkeys (*Macaca fuscata fuscata*). *Primates,* **28,** 343–352.

Johnson, D. B. (1982). Altruistic behavior and the development of the self in infants. *Merrill-Palmer Quarterly,* **28,** 379–388.

Johnson, M. H., & Morton, J. (1991). *Biology and cognitive development: The case of face recognition.* Cambridge, MA: Blackwell.

Jolly, A. (1964a). Choice of cue in prosimian learning. *Animal Behaviour,* **12,** 571–577.

Jolly, A. (1964b). Prosimians' manipulation of simple object problems. *Animal Behaviour,* **12,** 560–570.

Kamil, A. C. (1978). Systematic foraging by a nectar-feeding bird, the Amakihi (*Loxops vivens*). *Journal of Comparative and Physiological Psychology,* **92,** 388–396.

Kamil, A. C. (1983). Optimal foraging theory and the psychology of learning. *American Zoologist,* **23,** 291–302.

Kamil, A. C. (1984). Adaptation and cognition: Knowing what comes naturally. In H. L. Roitblat, T. G. Bever, & H. S. Terrace (Eds.), *Animal cognition.* Hillsdale, NJ: Erlbaum.

Kamil, A. C., & Yoerg, S. I. (1982). Learning and foraging behavior. In P. P. G. Bateson & P. H. Klopfer (Eds.), *Perspectives in ethology* (Vol. **5**). New York: Plenum.

Karniol, R. (1978). Children's use of intention cues in evaluating behavior. *Psychological Bulletin,* **85,** 76–85.

King, M. (1971). The development of some intention concepts in young children. *Child Development,* **42,** 1145–1152.

Köhler, W. (1927). *The mentality of apes.* London: Routledge & Kegan Paul.

Kummer, H. (1982). Social knowledge in free-ranging primates. In D. R. Griffin (Ed.), *Animal mind—human mind.* New York: Springer.

Lempers, J. D., Flavell, E. R., & Flavell, J. H. (1977). The development in very young children of tacit knowledge concerning visual perception. *Genetic Psychology Monographs,* **95,** 3–53.

Lewis, D. (1969). *Convention: A philosophical study.* Cambridge, MA: Harvard University Press.

Lewis, M., Sullivan, M. W., Stanger, C., & Weiss, M. (1989). Self-development and self-conscious emotions. *Child Development,* **60,** 146–156.

Liben, L. S. (1978). Perspective-taking skills in young children: Seeing the world through rose-colored glasses. *Developmental Psychology,* **14,** 87–92.

Lillard, A. S. (1993). Pretend play skills and the child's theory of mind. *Child Development,* **64,** 348–371.

Lockard, R. B. (1971). Reflections on the fall of comparative psychology: Is there a message for us all? *American Psychologist,* **26,** 168–179.

Lyon, T. D. (1993). *Young children's understanding of desire and knowledge.* Unpublished doctoral dissertation, Stanford University.

Macphail, E. M. (1982). *Brain and intelligence in vertebrates.* Oxford: Clarendon.

Macphail, E. M. (1987). The comparative psychology of intelligence. *Behavioral and Brain Sciences,* **10,** 645–656.

Marks, J. (1991). What's old and new in molecular phylogenetics. *American Journal of Physical Anthropology,* **85,** 207–219.

Marks, J. (1992). Genetic relationships among the apes and humans. *Current Opinion in Genetics and Development,* **2,** 883–889.

Masangkay, Z. S., McKluskey, K. A., McIntyre, C. W., Sims-Knight, J., Vaughn, B. E., & Flavell, J. H. (1974). The early development of inferences about the visual precepts of others. *Child Development,* **45,** 357–366.

Mathieu, M., & Bergeron, G. (1981). Piagetian assessment on cognitive development in chimpanzees (*Pan troglodytes*). In A. B. Chiarelli & R. S. Corruccini (Eds.), *Primate behavior and sociobiology.* New York: Springer.

McKinney, M., & McNamara, K. J. (1991). *Heterochrony: The evolution of ontogeny.* New York: Plenum.

Menzel, E. W., Jr. (1971). Communication about the environment in a group of young chimpanzees. *Folia Primatologica,* **15,** 220–232.

Menzel, E. W., Jr. (1974). A group of young chimpanzees in a one-acre field. In A. Schrier & F. Stollnitz (Eds.), *Behavior of non-human primates: Modern research trends.* New York: Academic.

Menzel, E. W., Jr., & Johnson, M. K. (1976). Communication and cognitive organization in humans and other animals. *Annals of the New York Academy of Sciences,* **280,** 131–146.

Mignault, C. (1985). Transition between sensorimotor and symbolic activities in nursery-reared chimpanzees (*Pan troglodytes*). *Journal of Human Evolution,* **14,** 747–758.

Miller, P. H., & Aloise, P. A. (1989). Young children's understanding of the psychological causes of behavior: A review. *Child Development,* **60,** 257–285.

Moses, L. J., & Chandler, M. J. (1992). Traveler's guide to children's theories of mind. *Psychological Inquiry,* **3,** 286–301.

Mossler, D. G., Marvin, R. S., & Greenberg, M. T. (1976). Conceptual perspective taking in 2- to 6-year old children. *Developmental Psychology,* **12,** 85–86.

O'Neill, D. K., & Gopnik, A. (1991). Young children's ability to identify the sources of their beliefs. *Developmental Psychology,* **27,** 390–397.

Parker, S. T. (1990). Origins of comparative developmental evolutionary studies of primate mental abilities. In S. T. Parker & K. R. Gibson (Eds.), *"Language" and intelligence in monkeys and apes.* New York: Cambridge University Press.

Parker, S. T., & Gibson, K. R. (1979). A developmental model for the evolution of language and intelligence in early hominids. *Behavioral and Brain Sciences,* **2,** 367–408.

Parker, S. T., & Gibson, K. R. (Eds.). (1990). *"Language" and intelligence in monkeys and apes.* New York: Cambridge University Press.

Perner, J. (1991). *Understanding the representational mind.* Cambridge, MA: MIT Press.

Perner, J., & Ogden, J. (1988). Knowledge for hunger: Children's problems with representation in imputing mental states. *Cognition,* **29,** 47–61.

Phillips, W., Baron-Cohen, S., & Rutter, M. (1992). The role of eye contact in goal detection: Evidence from normal infants and children with autism or mental handicap. *Development and Psychopathology,* **4,** 375–383.

Piaget, J. (1932). *The moral judgement of the child.* New York: Free Press.

Piaget, J., & Inhelder, B. (1956). *The child's conception of space.* London: Routledge & Kegan Paul.

Pillow, B. H. (1989). Early understanding of perception as a source of knowledge. *Journal of Experimental Child Psychology,* **47,** 116–129.

Potì, P., & Spinozzi, G. (1994). Early sensorimotor development in chimpanzees (*Pan troglodytes*). *Journal of Comparative Psychology,* **108,** 93–103.

Povinelli, D. J. (1991). *Social intelligence in monkeys and apes.* Unpublished doctoral dissertation, Yale University.

Povinelli, D. J. (1993). Reconstructing the evolution of mind. *American Psychologist,* **48,** 493–509.

Povinelli, D. J. (1994a). Comparative studies of mental state attribution: A reply to Heyes. *Animal Behaviour,* **48,** 239–241.

Povinelli, D. J. (1994b). How to create self-recognizing gorillas (but don't try it on macaques). In S. Parker, R. Mitchell, & M. Boccia (Eds.), *Self-awareness in animals and humans.* Cambridge: Cambridge University Press.

Povinelli, D. J. (1996). Chimpanzee theory of mind? The long road to strong inference. In P. Carruthers & P. Smith (Eds.), *Theories of theories of mind.* Cambridge: Cambridge University Press.

Povinelli, D. J., & Davis, D. R. (1994). Differences between chimpanzees (*Pan troglodytes*) and humans (*Homo sapiens*) in the resting state of the index finger: Implications for pointing. *Journal of Comparative Psychology,* **108,** 134–139.

Povinelli, D. J., & deBlois, S. (1992). Young children's (*Homo sapiens*) understanding of knowledge formation in themselves and others. *Journal of Comparative Psychology,* **106,** 228–238.

Povinelli, D. J., & Eddy, T. J. (in press). Chimpanzees: Joint visual attention. *Psychological Science.*

Povinelli, D. J., Nelson, K. E., & Boysen, S. T. (1990). Inferences about guessing and knowing by chimpanzees (*Pan troglodytes*). *Journal of Comparative Psychology,* **104,** 203–210.

Povinelli, D. J., Nelson, K. E., & Boysen, S. T. (1992). Comprehension of role reversal by chimpanzees: Evidence of empathy? *Animal Behaviour,* **43,** 633–640.

Povinelli, D. J., Parks, K. A., & Novak, M. A. (1991). Do rhesus monkeys (*Macaca mulatta*) attribute knowledge and ignorance to others? *Journal of Comparative Psychology,* **105,** 318–325.

Povinelli, D. J., Parks, K. A., & Novak, M. A. (1992). Role reversal by rhesus monkeys, but no evidence of empathy. *Animal Behaviour,* **44,** 269–281.

Povinelli, D. J., Perilloux, H. K., Reaux, J. E., & Bierschwale, D. T. (1995). *Young and juvenile chimpanzees' (*Pan troglodytes*) reactions to intentional versus accidental and inadvertent actions.* Unpublished manuscript.

Povinelli, D. J., Rulf, A. B., & Bierschwale, D. T. (1994). Absence of knowledge attribution and self-recognition in young chimpanzees (*Pan troglodytes*). *Journal of Comparative Psychology,* **108,** 74–80.

Povinelli, D. J., Rulf, A. B., Landau, K. R., & Bierschwale, D. T. (1993). Self-recognition in chimpanzees (*Pan troglodytes*): Distribution, ontogeny, and patterns of emergence. *Journal of Comparative Psychology,* **107,** 347–372.

Pratt, C., & Bryant, P. (1990). Young children understand that looking leads to knowing (so long as they are looking into a single barrel). *Child Development,* **61,** 973–982.

Premack, D. (1984). Pedagogy and aesthetics as sources of culture. In M. S. Gazzaniga (Ed.), *Handbook of cognitive neuroscience.* New York: Plenum.

Premack, D. (1988). "Does the chimpanzee have a theory of mind" revisited. In R. Byrne & A. Whiten (Eds.), *Machiavellian intelligence.* New York: Oxford University Press.

Premack, D., & Dasser, V. (1991). Perceptual origins and conceptual evidence for theory of mind in apes and children. In A. Whiten (Ed.), *Natural theories of mind.* Oxford: Blackwell.

Premack, D., & Woodruff, G. (1978a). Author's response. *Behavioral and Brain Sciences,* **1,** 616–629.

Premack, D., & Woodruff, G. (1978b). Does the chimpanzee have a theory of mind? *Behavioral and Brain Sciences,* **1,** 515–526.

Preuss, T. M. (1993). The role of the neurosciences in primate evolutionary biology: Historical commentary and prospectus. In R. D. E. MacPhee (Ed.), *Primates and their relatives in phylogenetic perspective.* New York: Plenum.

Preuss, T. M. (1995). The argument from animals to humans in cognitive neuroscience. In M. S. Gazzaniga (Ed.), *The cognitive neurosciences.* Cambridge, MA: MIT Press.

Preuss, T. M., & Goldman-Rakic, P. S. (1991a). Ipsilateral cortical connections of granular frontal cortex in the strepsirhine primate *Galago,* with comparative comments on anthropoid primates. *Journal of Comparative Neurology,* **310,** 507–549.

Preuss, T. M., & Goldman-Rakic, P. S. (1991b). Myelo- and cytoarchitecture of the granular frontal cortex and surrounding regions in the strepsirhine primate *Galago* and the anthropoid primate *Macaca. Journal of Comparative Neurology,* **310,** 429–474.

Ristau, C. A. (1991). Before mindreading: Attention, purposes and deception in birds? In A. Whiten (Ed.), *Natural theories of mind: Evolution, development and simulation of everyday mindreading.* Cambridge: Blackwell.

Robinson, P. W., & Foster, D. F. (1979). *Experimental psychology: A small-N approach.* New York: Harper & Row.

Ruffman, T. K., & Olson, D. R. (1989). Children's ascriptions of knowledge to others. *Developmental Psychology,* **25,** 601–606.

Rumbaugh, D. M., & Pate, J. L. (1984). The evolution of cognition in primates: A comparative perspective. In H. L. Roitblat, T. G. Bever, & H. S. Terrace (Eds.), *Animal cognition.* Hillsdale, NJ: Erlbaum.

Savage-Rumbaugh, E. S., Murphy, J., Sevcik, R. A., Brakke, K. E., Williams, S. L., & Rumbaugh, D. M. (1993). Language comprehension in ape and child. *Monographs of the Society for Research in Child Development,* **58**(3–4, Serial No. 233).

Scaife, M., & Bruner, J. (1975). The capacity for joint visual attention in the infant. *Nature,* **253,** 265–266.

Seligman, M. E. P. (1970). On the generality of the laws of learning. *Psychological Review,* **77,** 406–418.

Shea, B. T. (1983). Paedomorphosis and neoteny in the pygmy chimpanzee. *Science,* **222,** 521–522.

Shea, B. T. (1988). Heterochrony in primates. In M. L. McKinney (Ed.), *Heterochrony in evolution: A multidisciplinary approach.* New York: Plenum.

Shettleworth, S. J. (1975). Reinforcement and the organization of behavior in golden hamsters: Hunger, environment, and food reinforcement. *Journal of Experimental Psychology: Animal Behavior Processes,* **1,** 56–87.

Shultz, T. R., & Shamash, F. (1981). The child's conception of intending act and consequence. *Canadian Journal of Behavioral Science,* **13,** 368–372.

Shultz, T. R., Wells, D., & Sarda, M. (1980). Development of the ability to distinguish intended actions from mistakes, reflexes, and passive movements. *British Journal of Social and Clinical Psychology,* **19,** 301–310.

Smith, M. C. (1978). Cognizing the behavior stream: The recognition of intentional action. *Child Development,* **49,** 736–743.

Smuts, B. (1985). *Sex and friendship in baboons.* New York: Aldine.

Spinozzi, G. (1993). Development of spontaneous classificatory behavior in chimpanzees (*Pan troglodytes*). *Journal of Comparative Psychology,* **107,** 193–200.

Spitz, R. A. (1965). *The first year of life.* New York: International Universities Press.

Spitz, R. A., & Wolf, K. M. (1946). The smiling response: A contribution to the ontogenesis of social relationships. *Genetic Psychology Monographs,* **34,** 57–125.

Tomasello, M., & Call, J. (1994). The social cognition of monkeys and apes. *Yearbook of Physical Anthropology,* **37,** 273–305.

Tomasello, M., Call, J., Nagell, K., Olguin, K., & Carpenter, M. (1994). The learning and use of gestural signals by young chimpanzees: A trans-generational study. *Primates,* **35,** 137–154.

Tomasello, M., Kruger, A. C., & Ratner, H. H. (1993). Cultural learning. *Behavioral and Brain Sciences,* **16,** 495–552.

Trevarthen, C., & Hubley, P. (1978). Secondary intersubjectivity: Confidence, confiders, and acts of meaning in the first year. In A. Lock (Ed.), *Before speech: Beginning of interpersonal communication.* London: Academic.

Umiker-Sebeok, J., & Sebeok, T. A. (1980). Introduction: Questioning apes. In T. A. Sebeok & J. Umiker-Sebeok (Eds.), *Speaking of apes: A critical anthology of two-way communication with man.* New York: Plenum.

van Schaik, C. P., van Noordwijk, M. A., Warsono, B., & Sutriono, E. (1983). Party size and early detection of predators in Sumatran forest primates. *Primates,* **24,** 211–221.

Wasserman, E. A. (1981). Comparative psychology returns: A review of Hulse, Fowler, and Honig's "Cognitive processes in animal behavior." *Journal of the Experimental Analysis of Behavior,* **35,** 243–257.

Watts, D. P., & Pusey, A. E. (1993). Behavior of juvenile and adolescent great apes. In M. E. Pereira & L. Fairbanks (Eds.), *Socioecology of juvenile primates.* Oxford: Oxford University Press.

Wellman, H. M. (1990). *The child's theory of mind.* Cambridge, MA: Bradford.

Whiten, A. (Ed.). (1991). *Natural theories of mind: Evolution, development and simulation of everyday mindreading.* Cambridge: Blackwell.

Whiten, A. (1993). Evolving theories of mind: The nature of non-verbal mentalism in other primates. In S. Baron-Cohen, H. Tager-Flusberg, D. Cohen, & F. Volkmar (Eds.), *Understanding other minds.* Oxford: Oxford University Press.

Whiten, A., & Byrne, R. W. (1988). Tactical deception in primates. *Behavioral and Brain Sciences,* **11,** 233–244.

Wimmer, H., Hogrefe, G.-J., & Perner, J. (1988). Children's understanding of informational access as a source of knowledge. *Child Development,* **59,** 386–396.

Wimmer, H., & Perner, J. (1983). Beliefs about beliefs: Representation and constraining function of wrong beliefs in young children's understanding of deception. *Cognition,* **13,** 103–128.

Woodruff, G., & Premack, D. (1979). Intentional communication in the chimpanzee: The development of deception. *Cognition,* **7,** 333–362.

Woolley, J. D., & Wellman, H. M. (1993). Origin and truth: Young children's understanding of imaginary mental representations. *Child Development,* **64,** 1–17.

ACKNOWLEDGMENTS

The research and preparation of this manuscript were funded in part by National Institutes of Health grant RR-03583-05 to the New Iberia Research Center and National Science Foundation Young Investigator Award SBR-8458111 to Daniel J. Povinelli. We are deeply indebted to the director of the University of Southwestern Louisiana (USL)–New Iberia Research Center, Dr. William E. Greer, who provided space, personnel, and animal resources necessary to carry this project from inception to fruition. A special debt is due to Mr. Anthony Rideaux, who was responsible for overseeing the day-to-day training and caretaking of the subjects used in these investigations. A number of students participated in the discussions of the logic and design of the experiments as well as the training and testing of the children and chimpanzees: Danielle Bacqué, Donna Bierschwale, Gabrielle Hebert, Keli Landau, Helen Perilloux, Jim Reaux, Nicole Spencer, and Mia Zebouni. We thank Claude G. Čech, Steve Giambrone, and Anthony Maida for valuable discussions and comments on the manuscript. We thank Mr. John Hardcastle, Dr. Jeff Rowell, and Mr. John D. Boutte for their professional and technical support during the investigations with the chimpanzees. Photographs are by Donna Bierschwale. Finally, we are deeply grateful to the parents and children who agreed to participate in this research as well as the directors and staff of the USL Child Development Center, Lourde's Child Care Center, Kid's Korner, and the First Methodist Day Care Center. Correspondence should be addressed to Daniel J. Povinelli, Laboratory of Comparative Behavioral Biology, University of Southwestern Louisiana–New Iberia Research Center, 4401 W. Admiral Doyle Dr., New Iberia, LA 70560.

COMMENTARY

ON NOT UNDERSTANDING MINDS

R. Peter Hobson

Povinelli and Eddy pose the question, Do chimpanzees interpret seeing as a mental event or as an exclusively behavioral act? Elsewhere the word *interpret* is replaced by *understand,* as in the following account of the authors' theoretical strategy: "In order to assess whether young chimpanzees possess an understanding that visual perception connects organisms to the external world, we first constructed two theoretical frameworks (behaviorist vs. mentalistic) to explain their behavior" (p. 25).

This strategy has paid off handsomely. Through a series of carefully constructed experiments, the investigators provide persuasive evidence against one of the two theoretical options they have offered us. We should not suppose that chimpanzees are like adult human beings in understanding that eye gaze is accompanied by a subjective registration of events in the world. As the authors summarize: "Collectively, our findings provide little evidence that young chimpanzees understand seeing as a mental event" (p. vi). Thus, the authors contribute to a tradition of research in which investigators have attempted to tease out whether careful analyses of the behavioral contingencies that apply in a given situation might enable an organism that has little or no awareness of mental states to predict other creatures' behavior and thereby to simulate psychological understanding. Such circumspection is vitally important if we are not to lapse into anthropomorphism and uncritically allow that any number of organisms that behave like human beings have human-like sophistication in their understanding of minds. Having acknowledged this, I should like to explore some potential hazards in formulating opposing and incompatible theoretical positions in this way. In particular, there is a danger that by dichotomizing the theoretical alternatives, even as a "first heuristic approximation" (p. 132), one might presuppose that if either one of these alternatives is incorrect, the alternative must be correct. When we consider the evolution of psychological understanding, however, the dichotomy itself may be only partly valid. In note 1 (see p. 25), Povinelli and Eddy emphasize how the theoretical frameworks they set out lie at the extreme ends on a spectrum of views. So perhaps my own remarks amount to little more than a gloss on a note!

In this Commentary, then, I shall consider some complexities in interpreting experiments such as these, when we are attempting to evolve a perspective on the phylogenesis and ontogenesis of so-called theory of mind. I shall not focus on issues of experimental methodology, except insofar as these shape the way we think about the elements of social understanding in chimpanzees and human infants. I hope that such an approach is appropriate, given the statement by Povinelli and Eddy that their "ultimate aim is to reconstruct the timing and order of the evolution of various aspects of metacognition" (p. 2). I am concerned that the authors' theoretical approach might lead one to suppose that a purely behaviorist account is appropriate for animals up to a certain point in evolution and that beyond this one would need to attribute to animals a mentalistic view in order to explain their behavior toward others. In this case, however, an obvious question would present itself: How does this abrupt transition occur in phylogenesis? If there were *nothing* mentalistic in the organism's apprehension of other organisms prior to the advent of mental understanding, how is it that such understanding is ever achieved—and, moreover, achieved in such a way that mental events are understood to be so closely linked to behavioral events? This is also a question for those who wish to account for the ontogenesis of interpersonal understanding in young children (Hobson, 1991).

Consider again how Povinelli and Eddy present the two theoretical options. (They also consider a third option, in which organisms recognize that vision can lead to knowledge, but I shall not dwell on this.) We find that, in addition to offering the formulations I cited at the beginning of this Commentary, the authors contrast the possibility that organisms merely respond to other organisms' eyes, without appreciating the functional significance of those eyes, with the possibility that organisms can understand that vision subjectively links individuals to the world. Concerning the second of these alternatives, they explain: "At this level of information processing, visual perception may be equated with a subjective connection to the external world" (p. 18). In fact, of course, if terms such as *vision* or *visual perception* were appropriate for characterizing an aspect of the information processed, this would already imply that the organism had understood something about the subjective dimension behind another creature's bodily appearances. Therefore, such "equating" would really mean either (*a*) that the organism perceived and/or conceived of a subjective orientation *in* the other creature's eye-face orientation or (*b*) that the organism perceived and/or conceived that subjectivity was a property of the other organism and then inferred that the other's eye-face position was relevant for judgments about the content ("aboutness") of the other's experiences. This distinction may seem to be a trivial one, but I think that the research of Povinelli and Eddy illustrates its significance. Moreover, I hope that this way of putting things may clarify how there are at least three distinct parts to the opposed theoretical positions: the first has to do with the epistemological status of the perception or apprehension or registration or reaction or re-presentation or understanding or knowledge that we attribute to chimpanzees; the second has to do with the content of such awareness, especially its depth and sophistication in relation to the subjectivity and intentionality of individuals' mental states; and the third has to do with the specifics of "to what" (eyes or faces or bodies or actions or centers of consciousness or individuals or selves) such subjective or mentalistic status may be ascribed. For example, to represent the fact that eye direction and subjectivity are linked involves a far more articulated level of understanding than that involved in merely reacting to an individual's bodily expressive orientation. The appropriateness of each of these possible characterizations of social perception and understanding at each stage in a chimpanzee's or child's development, and the developmental continuities and discontinuities among the psychological abilities so characterized, are critical for the issues under discussion.

Let me try to flesh out what I mean. The evidence from the present studies suggests that, in settings where an observed individual's "looking" is abstracted from other aspects of his or her behavior, chimpanzees do not react to line of regard per se as if it were backed by subjective experience. Povinelli and Eddy demonstrate that there are circumstances in which young chimpanzees spontaneously attend to and follow the visual gaze of another individual, without reacting to such gaze as an expression of the other's psychological engagement. Therefore, if there were other situations in which the evidence seemed to suggest that chimpanzees have greater awareness of the nature of vision, we would hesitate before concluding either (*a*) that their appreciation of psychological meaning in eye gaze is generalizable and, especially, that chimpanzees can invoke something like concepts that represent vision *as* vision or (*b*) that such apprehension of meaning arises out of some mechanism that is specifically related to eye gaze, when it now appears more plausible that eye/face gaze following may initially occur without much psychological awareness. Insofar as it is justified to extrapolate to related phenomena in human infant development, these considerations are more in keeping with the accounts offered by Moore (1994) and Hobson (1994) than with the contrasting approach of Baron-Cohen (1994).

For anyone tackling these problems, it is difficult to maintain the (relative) separateness of two issues: first, whether and under what conditions a chimpanzee or infant relates only to the behavior or also to the mind of another, or perhaps at times to bodily anchored meanings that are in a specific sense both behavioral and mental; and, second, whether the nature of such social relatedness should be characterized in terms of capabilities for direct perception, or implicit awareness of behavioral patterns and/or mental states, or as reflecting the possession of more or less full-blown concepts about the minds or mental faculties (e.g., seeing) of others. The evidence from the present studies suggests that, if one tests chimpanzees in circumstances when they are observing "eyes" relatively abstracted from facial and other emotional expressions and actions, then they do not respond like mature human beings. It seems reasonable to conclude that chimpanzees do not have direct perceptual capacities for apprehending subjective states in the eyes alone and that they do not have concepts about the mental life of others that they can deploy when interpreting the directedness of others' eyes. Therefore, it is unlikely that eye gaze following per se is sufficient to ground mental state awareness, at however basic or implicit a level this is considered. I think we have learned something important both about chimpanzees' potential for reacting to "nonsubjective" but nevertheless psychologically relevant aspects of behavior, in this case a form of directedness that is associated with eye-face position (studied in children by Butterworth & Jarrett, 1991), and about chimpanzees' lack of concepts of mind that could be applied to fill out the psychological meaning of directed eye gaze.

Why, then, did I begin by acknowledging the heuristic value of dichotomizing behaviorist and mentalist views on chimpanzees' understanding but then proceed to express unease about the distinction? The principal reason is that in order to understand the evolution of metacognition, both across species and in the course of early human development, we may need to adopt a vocabulary in which the vital distinctions between behavior and mind, and between percepts and concepts, are both respected and transcended. To begin with, we need to capture how chimpanzees and human infants may react to more than behavior, if behavior is narrowly conceived to contrast with expressions of subjective mental life. For example, chimpanzees may perceive and react to attitudes as manifest in behavioral expressions and actions, without yet understanding what it is they are reacting to. They might respond emotionally to other primates' expressions of feeling (not tested here) and even use information from the other's behavior and directedness of eye gaze when reacting to (what we would call) the focus of the other's attention, without recognizing the latter *as* a focus of the other's attention. All this might be achieved without chimpanzees *understanding* anything about the subjectivity and intentionality (including the directedness) of mental states and, specifically, without chimpanzees understanding the eyes as subjectively connecting other primates with the world (however *the world* is itself understood). On the other hand, one could mount an argument that, under these circumstances, the chimpanzee has demonstrated how it can register a relationship between something other than "exclusively behavioral acts" and the objects and events toward which those acts are directed. Surely the chimpanzee would be reacting to what we human beings recognize to be behaviorally expressed and intentionally directed emotional *states* in the other, in such a manner that the chimpanzee's own relationship to the environment is altered in particular ways. It might be registering certain of the implications of what we would describe as a subjective orientation, without representing to itself that the observed individual has a subjective orientation. The possibility arises that such non-theory-like relations between the perceiving organism and others are foundational for what in human beings becomes an explicit understanding of the partially separable components of behavior, on the one hand, and mind, on the other.

This brings us to the dichotomy between percepts and concepts. Both chimpanzees' and human infants' level of understanding may fall far short of having concepts about the nature of mental life, but not so far short that their perception of others is on the level of our perception of robots. I shall not attempt to analyze what it would mean for chimpanzees to have concepts per se, but I should like to make one observation about concepts concerned with mental life. As in other conceptual domains, we are dealing with a nexus of concepts that link with and support each other. Progress in understanding minds entails and is entailed by developments in understanding how oneself and other embodied people are "selves" to whom minds may be ascribed. An individual who understands minds is understanding both the commonality among and the differentiation of individuals qua centers of subjectivity. It is likely that to perceive actions or to respond to bodily expressions of feeling is a developmental and epistemological precondition for understanding the nature of agents and subjects of experience; I believe that to perceive the emotional quality and directedness of mental states is a developmental and epistemological precondition for understanding how given objects and events can have different meanings or fall under different descriptions for different individuals (the intentionality of mental states). It may yet prove to be the case that chimpanzee-level sensitivity to gaze directedness may contribute to, even though it is insufficient for, the phylogenesis and ontogenesis of understanding about the intentionality of mind. Yet, in order to "represent the facts" of subjective mental life in such a way as to interpret elements of behavior (e.g., the functional significance of the eyes) in terms of those facts, a great deal needs to be understood about oneself as a self among others, existing in a world that may be construed in different ways by different individuals.

I would argue that in certain important respects, therefore, our theoretical dichotomy splits asunder what nature has given to us as an integrated whole. Povinelli and Eddy protest that "at present we have no way of teasing apart the emotional valence of such mutual gaze (in which we were less interested) from its cognitive interpretation (in which we were most interested)" (p. 127). This methodological limitation should not become a theoretical limitation. Povinelli and Eddy constructed tests in which chimpanzees were faced with (*a*) cool, relatively nonemotional people, so that affective information was reduced to a minimum, and (*b*) novel tasks in which the chimpanzees had to recombine existing actions and in so doing apply relatively abstract forms of understanding. Sure enough, 2-year-old children could apply an interpretative framework that was not available to chimpanzees, although we must not forget that 2-year-olds have a powerful *form* of conceptual/linguistic framework that dramatically increases the flexibility and generalizability with which they can apply their understandings. All this is perfectly in order for demonstrating that chimpanzees do not have certain concepts already acquired by 2-year-old humans. As we have seen, the approach has also proved very successful in elucidating potentially dissociable components of social perception and/or understanding. What we should not presume is that cognitive interpretations of subjective mental life are dissociable from emotional interpretations when considered over the course of evolution or over individuals' development. From the perspective of genetic epistemology, the latter kind of social awareness may provide necessary foundations for the former. If we confine ourselves to testing "cognitive interpretations," we may overlook critical transitions from affectively grounded, nonconceptual forms of social interpretation to more cognitively abstracted modes of understanding.

Finally, then, it is worth reflecting on what Povinelli and Eddy call their ultimate aim, namely, to reconstruct the timing and order of evolution of various aspects of metacognition and to do so through consideration of what distinguishes chimpanzees and most other primates from developments in humans. If we accept (as I do) that "young chimpanzees possess and learn rules about visual perception" but that "these rules do not necessarily incorporate the notion that seeing is 'about' something" (p. vi), how could such a notion of mental aboutness have become an attainable concept for us? How do human beings acquire such a notion in ontogenesis? I have already alluded to the fact that, to acquire such a concept, one needs to acquire associated concepts to do with individual selves who have outwardly focused attitudes that are at once like one's own attitudes but also distinctive to each individual. My hunch is that what distinguishes chimpanzees from humans (and also autistic from nonautistic children) is the degree to which one individual is drawn into an engagement and identification *with* the mental states of another. It is one thing to react to the emotional expressions or actions of another, but it involves something more if one moves to the state or adopts the actions of the other and, a fortiori, if one is aware of being engaged in something like sharing experiences or empathizing or role taking. My guess is that a relatively small increment in this set of propensities—propensities that need to be preconceptual and biologically programmed, not at first the *result* of advances in interpersonal understanding—is the mechanism by which human beings acquire a greater degree of anchorage in the psychological perspective of the other. I have suggested elsewhere that it is through such anchorage that a cognitive separation of self from other, of attitude from object of attitude, and ultimately of symbolic representations from their referents, is achieved (Hobson, 1993). On the one hand, therefore, we need to review evidence that chimpanzees may show greater sensitivity to bodily expressed states involving looking that convey more vividly how the other is psychologically/emotionally engaged with the world; and, on the other hand, we need to examine closely whether such sensitivity ever involves an orientation specifically to the other's outer-directed state (rather than the environment related to), in the way that can occur in human infants from around the end of their first year of life. The accompanying Commentary by Michael Tomasello provides an excellent starting point for this appraisal.

I would like to end by emphasizing that I have said little if anything that directly contradicts the account provided by Povinelli and Eddy. These investigators have explored whether chimpanzees understand the role of the eyes in establishing attentional focus—and their results suggest that chimpanzees do not perceive "eyes" per se (or, indeed, relatively expressionless people with eyes) as windows on minds. They have also provided valuable evidence against the view that chimpanzees conceptualize other beings as having their own subjective mental states, insofar as they do not respond to others' eye gaze in experimental settings as they might be expected to do if they had such concepts. Having said this, we need to hesitate before characterizing chimpanzees' registration of all looking behavior as "exclusively behavioral." There remains a great deal of room for chimpanzees to have the potential for nonbehaviorist modes of interpreting the vision-related expressions and behavior of others. We also need to be careful in weighing the epistemological status of different forms of "interpreting," and I am uneasy about the freedom with which Povinelli and Eddy employ such terms as *know, representing, understanding,* and *conception.* The range of chimpanzees' and human infants' preconceptual modes of relating to others, especially those ways of relating that escape definition in either behaviorist or mentalistic terms, may yet prove to be pivotal for our understanding of the evolution of increasingly elaborate forms of "knowing" and specifically for our account of the evolution and ontogenesis of metacognition.

References

Baron-Cohen, S. (1994). How to build a baby that can read minds: Cognitive mechanisms in mindreading. *Current Psychology of Cognition,* **13,** 513–552.

Butterworth, G., & Jarrett, N. (1991). What minds have in common is space: Spatial mechanisms serving joint visual attention in infancy. *British Journal of Developmental Psychology,* **9,** 55–72.

Hobson, R. P. (1991). Against the theory of "theory of mind." *British Journal of Developmental Psychology,* **9,** 33–51.

Hobson, R. P. (1993). *Autism and the development of mind.* Hove: Erlbaum.

Hobson, R. P. (1994). Perceiving attitudes, conceiving minds. In C. Lewis & P. Mitchell (Eds.), *Children's early understanding of mind: Origins and development.* Hove: Erlbaum.

Moore, C. (1994). Intentionality and self-other equivalence in early mindreading: The eyes do not have it. *Current Psychology of Cognition,* **13,** 661–668.

COMMENTARY

CHIMPANZEE SOCIAL COGNITION

Michael Tomasello

In their natural social interactions chimpanzees do many things that would seem to involve one individual monitoring the visual attention of another. For example, in their natural gesturing with conspecifics chimpanzees use visually based gestures to solicit play (e.g., "arm raise") only if the potential recipient is facing them. When they wish to solicit play from individuals whose backs are turned, they typically poke them, or throw something at them, or produce some attention-getting gesture such as slapping the ground noisily (to get others to look at their "play face"). When adult chimpanzees wish to touch a neonate or to obtain some food close to a dominant individual, they know to attempt these things only when the mother's or dominant's back is turned (Tomasello, Call, Nagell, Olguin, & Carpenter, 1994).

The findings of Povinelli and Eddy will thus be surprising to most researchers who have worked extensively with apes. It seems that their seven chimpanzees, over half the way to sexual maturity and with much experience interacting with humans, know very little about how human eyes work. For example, they used a visually based gesture to beg from a human just as frequently when she was wearing a blindfold over her eyes as when she was wearing a blindfold over her mouth or just as frequently when she was staring at the ceiling as when she was looking toward their gesture. The chimpanzees clearly did differentiate a human whose back was turned from a human who was facing toward them, however, so they do have *some* knowledge of the conditions necessary for their begging gesture to achieve its intended goal—they just do not comprehend the precise role of the eyes

I would like to thank Josep Call and Ann Kruger for comments on this Commentary.

in the process. What this means is that many of the previous observations of apes monitoring the visual gaze of others must be reinterpreted in terms of something like whole body orientation or the like.

But Povinelli and Eddy want to know more than just whether chimpanzees know about the eyes as physical mechanisms. They want to know whether chimpanzees understand that there is someone behind those eyes, choosing to pay attention to some things but not to others and adopting one or another course of action as a result. That is, they want to know whether chimpanzees understand the "mental significance" of the eyes as the instruments of intentional beings. Unfortunately, the current studies are not definitive on the issue. Thus, on the one hand, the studies suggest that chimpanzees do not understand much about others as intentional beings. How could they if they actively gestured to an experimenter who had a bucket over her head? On the other hand, however, the 2½–3-year-old children whom Povinelli and Eddy tested as a comparison group frequently gestured to a human with a piece of cardboard covering her face, and these are beings who presumably have some understanding of others as intentional beings—after all, they have been having conversations with adults for over a year, adjusting their conversations to the adult's knowledge states, self-correcting when the adult does not understand, using some mental state terms such as see and want, and, in general, seeming to treat others as intentional agents (Akhtar & Tomasello, in press). The implication is that organisms, including Povinelli and Eddy's chimpanzees, may understand that others make psychological contact with the world without understanding the workings of the physical mechanisms, such as eyes, by means of which they do so—the recent speculations of Baron-Cohen (1995) notwithstanding. It is also possible that Povinelli and Eddy's chimpanzees are just too young to demonstrate adult understanding of the eyes or that they have an understanding of eyes only when they are interacting with conspecifics.

If an understanding of the workings of the eyes is not necessary for an understanding of intentionality—and if there might be something special about Povinelli and Eddy's subjects or procedures—the question remains, Do chimpanzees understand others as intentional beings? With some leanings in the skeptical direction, Povinelli and Eddy express a generally open mind on this question. But they do so without reviewing other lines of evidence in a systematic way—quite appropriately, as they are focused on explaining their own results. Consequently, although I cannot provide a thorough review in the limited space available here, I would like to provide a brief overview of some other lines of research that have a bearing on this issue—as a way of seeing how the current results fit into the overall picture emerging from recent studies of primate social cognition (for more thorough reviews, see Tomasello, in press; Tomasello & Call, 1994).

Experimental Evidence

There are two main experimental studies that purport to show that chimpanzees understand others as intentional beings. The first is that of Premack and Woodruff (1978), who had the language-trained chimpanzee Sarah choose pictures to complete video sequences of intentional human actions (e.g., she had to choose a picture of a key when the human in the video could not exit a locked door). This study was thoroughly criticized by Savage-Rumbaugh, Rumbaugh, and Boysen (1978a), however, who produced similar results with two other chimpanzees through purely association learning processes; for example, their apes also chose a picture of a key when shown only a picture of a lock *with no human action occurring at all* (they also provided alternative explanations of Sarah's behavior in various control conditions). Premack (1986) also reported briefly that, in a subsequent study, he could not train Sarah to discriminate between videos of humans engaged in intentional and nonintentional actions, and Povinelli (in ongoing work with Perilloux cited in the *Monograph*) has not been able to find this discrimination either.

The other main study is that of Povinelli, Nelson, and Boysen (1990). These investigators found that three of four chimpanzees preferred to ask for food from a person who had witnessed its hiding over someone who had not witnessed its hiding—the inference being that they could discriminate a "knowledgeable" from an "ignorant" human. The problem is that the apes in this study learned to do this only over many scores of trials with feedback on their accuracy after every trial (for details, see Heyes, 1993; Povinelli, 1994). This criticism also applies to the study of Woodruff and Premack (1979), in which chimpanzees learned to direct humans to the box without food (so they could obtain the one with food—what some call "deception") after many trials with feedback. Indeed, Povinelli and Eddy are acutely aware of this problem in the research reported here, and in many cases they explicitly use information on the subjects' patterns of learning over trials to help them decide whether an individual brought a piece of knowledge to the experiment or simply learned some specific cue during the experiment.

A related line of experimental evidence concerns chimpanzee social learning (for reviews, see Tomasello, 1990, 1994b, in press-a). For example, my collaborators and I have found that when apes observe others use tools they seem to focus on the affordances of the tool and the changes of state in the environment that the tool brings about; they do not pay attention to the specific method of tool use demonstrated for them, even when such attention would be beneficial to acquisition of the food they so clearly desire (what we have called "emulation learning"; e.g., Call & Tomasello, 1995;

Nagell, Olguin, & Tomasello, 1993; Tomasello, Davis-Dasilva, Camak, & Bard, 1987). One interpretation of this result is that apes do not follow into the behavioral strategies of others because they do not comprehend the intentionality that is embodied in their methods of tool use. Chimpanzees also show no evidence of imitatively learning the intentional gestures of their conspecifics, presumably because they are not able to understand and follow into the intentional significance of the gestures of others (Tomasello et al., 1994; Tomasello, Gust, & Frost, 1989).

Naturalistic Evidence

It is quite possible that these experimental studies underestimate chimpanzee social cognitive skills since in these studies the subjects must, for the most part, interact with someone who is not a member of their own species. Three sets of observations of chimpanzees in more natural social circumstances would seem, at first glance at least, to lend some credence to this hypothesis. (1) Menzel (1971) found that when one chimpanzee was shown where food had been hidden in a large field others who were subsequently introduced into that field monitored the knowledgeable individual's movements and on this basis were able to find the food—even extrapolating her direction of travel to arrive at the food before she did. (2) Chimpanzees (as well as many other primates and mammals) form coalitions and alliances for purposes of intergroup competition that seem to show much social intelligence (Harcourt & de Waal, 1992). The coalitions and alliances of primates seem to be especially sophisticated as individuals actively recruit specific allies who can help them with specific opponents (e.g., an ally who outranks their opponent). (3) There have also been reported, in "anecdotal" form, a number of observations of primate social strategies that some researchers have interpreted as deceptive (one individual supposedly attempting to create a false belief in another; Whiten & Byrne, 1988).

The problem is that, while all three of these sets of observations clearly indicate that chimpanzees and other primates are skillful at predicting and even manipulating the behavior of conspecifics, they do not tell us the social-cognitive bases of those skills. First, there is no question that chimpanzees can learn to predict the behavior of others on the basis of both contextual and behavioral cues in Menzel-like situations. For example, the fact that others are excited (as clearly expressed in their facial expressions and body posture) may indicate the presence of an exciting food, and the direction of their travel may be learned as a cue to its location—on the basis of experience with others in similar situations in the past. There is no need in any of this, however, to read the intentional or mental states of others.

Second, the social strategies apparent in coalitions and alliances are used as active means to bring about desired situations, on the basis of an ability to predict the behavior of others in specific situations and indeed to actively bring about those situations. These strategies are made more complex by the fact that primates have much knowledge of the complex social fields in which they operate, consisting of (*a*) a recognition of individual group mates, (*b*) a knowledge of their own relationship to these group mates based on past interactions with them, and (*c*) an understanding of the relationships that other individuals have with one another based on *their* past interactions. This kind of sophisticated social knowledge and predictions and strategies based on this knowledge are at the heart of even the most mundane primate social interactions and, in general, are the core of what de Waal (1982) has called primate politics. But again they all may be seen to concern the *behavior* of others, who have certain behavioral relationships with other individuals, not their intentional or mental states.

Third, with regard to deception, interpretation is an especially difficult issue, as almost all the reported observations are anecdotes (for the view that "the plural of anecdote is not data," see Bernstein, 1988). But one seemingly reliable instance was reported by de Waal (1986), who on repeated occasions observed an individual chimpanzee hold out its hand to another in an apparent appeasement gesture but attack the other when he approached her. De Waal interpreted this behavior as deception: the perpetrator wanted the other to believe that she had friendly intentions when in fact she did not. It is just as likely, however, that the perpetrator wanted the other individual to approach her (so she could attack) and so performed a behavior that had in the past led to that result—in other contexts. This use of an established social behavior in a novel context is clearly a very intelligent strategy for manipulating the behavior of others, but it is not clear that it involves the understanding and manipulation of the intentional or mental states of others.

Obviously, reasonable people may differ in the richness with which they interpret these three sets of naturalistic observations; I have emphasized a "lean" interpretation only. However, if chimpanzees in their natural habitats understand conspecifics in terms of their intentional and mental relations to the world, not just in terms of predictable behavioral sequences, there are a number of things that we might expect them to do that they seemingly do not do. For example, in their natural habitats, chimpanzees do not spontaneously point for others to distal entities in their environments, either with or without index finger extension. They also do not hold objects up to show them to others, or try to bring others to locations so that they can observe things there, or actively offer objects to other individuals by holding them out to them. Chimpanzees and other apes also do not engage in the intentional teaching of offspring (the two most reliable observations that have been interpreted as ape teaching, e.g., by Boesch, 1991, have other, equally plausible, interpretations; Kruger & Tomasello, in press). And, finally, although chimpanzees do engage in group hunting in the wild (Boesch & Boesch, 1989), cooperation between individuals in experimental situations does not reveal sophisticated skills of directing others in desired ways to enhance the cooperative effort (Chalmeau, 1994; Crawford, 1937).

In my view, the situation is this. Predicting the behavior of others is one thing; understanding it as intentional is another. Predicting behavior means learning that one thing usually leads to another, and, in combination with a knowledge of the past behavior of others in various circumstances, it may lead to the formation of a variety of effective social strategies in which conditions are actively created so as to manipulate the behavior of others. Understanding behavior as intentional, on the other hand, means understanding behavior as an integrated process containing as separable components goals, behavior means, and perceptual monitoring: an intentional agent has different behavioral means to achieve a particular goal (i.e., has choices to make) and also has voluntary control of its attention (again, has choices to make) as it monitors its progress toward that goal. Understanding others in this way opens up the possibility of manipulating their behavior, not directly, but by doing something to affect either their goals or their perceptual monitoring of their progress toward that goal. But this is precisely what we do not see chimpanzees doing.

What about Human Infants?

In several places, Povinelli and Eddy imply that if a skeptical eye were turned on human children they also might not look so sophisticated in their skills of social cognition. The many studies in which 4-year-old children show clear evidence for understanding the mental life of others, including their false beliefs, would seem to refute this suggestion. But I would like to make the even stronger claim that there are a number of lines of evidence from much earlier in ontogeny—from around 12–18 months of age, in fact—that human children understand other persons as psychological beings with intentional relations to the world (Baressi & Moore, in press). Indeed, my claim is that it is in these early manifestations of human social cognition that we already see the difference in ape and human social cognition (Tomasello, 1995a, in press-c).

First of all, during their second year of life human infants are clearly inclined both to follow into and to direct the intentional states of others (not just their behavior). One clear example comes from studies of infant imitative learning of the intentional actions of others. In one study, 14-month-old infants observed an adult bend at the waist and touch her head to a panel, thus turning on a light (Meltzoff, 1988). The infants reproduced this unusual and awkward behavior, and they did so even though it would have been much easier and more natural for them simply to push the panel with their hand. By reproducing the adult "strategy" in this situation (not just her end result), it would seem that infants understood that (1) the adult had the goal (intention) of illuminating the light and then chose one means for doing so, from among other possible means, and (2) if they had the same goal (intention) they could choose the same means. (Carpenter, 1995, has reproduced these results with slightly younger children, with the additional requirement that they look in anticipation to the light as soon as they reproduce the demonstrator's behavior—to rule out the simple mimicking of body movements.) In a more recent study, Meltzoff (1995) found that when confronted with an adult engaged in an action on an object 18-month-old infants reproduced the action they saw adults *attempting* to perform, not the one she actually did perform. The difference with ape social learning is that the infants in both these studies were tuning in to the adult and her intentions in the situation, not just the changes of state in the world that her behavior brought about. (For evidence of gaze following by 1-year-old infants in an experimental situation different from Povinelli and Eddy's, see also Corkum & Moore, 1995.)

Infant skills at following into the intentional states of others are even more apparent in language-learning situations. For example, in a recent series of studies we had an adult announce to 18- and 24-month-old children, "Let's go find the modi" (Tomasello & Barton, 1994; Tomasello, Strosberg, & Akhtar, in press). Using no language after this, the adult then proceeded to a row of buckets and extracted several objects with a scowl on her face, then an object that brought her obvious glee. Almost all the children knew immediately that the first objects extracted were not the modi (the adult had scowled at them) and that the last object extracted was the modi (the adult had smiled at it). They knew this because they understood that the adult had an intention to find this thing called a modi and that all her subsequent behavior was organized around that goal. The frowns thus meant that those first objects were not the modi; the smile meant fulfillment of the intention and thus that the modi had been found. (Note that eye gaze direction is not diagnostic in this study—the rejected objects were actually looked at first.) Other recent studies have shown that children can learn words in a variety of other situations in which eye gaze direction is not diagnostic, suggesting that word learning is dependent in a fundamental way on a variety of flexible strategies for reading the communicative intentions of others (for a review, see Tomasello, in press-b).

In addition to these social learning strategies, from around their first birthdays human infants also make regular attempts to direct the attention of adults to outside entities through various forms of intentional communication. This includes the use of declarative gestures such as pointing and showing whose purpose is not to obtain things but only to share attention to them with other persons—as clearly demonstrated by the recent microanalytic studies of Franco and Butterworth (in press). As was noted above, apes do not as a matter of course perform declarative gestures of this type. And again the use of language makes this ability to manipulate intentional states even more apparent. Thus, at around 18 months of age children begin to engage in acts of predication; that is, in one communicative act they both establish a topic (either linguistically or nonlinguistically) and make some comment on it. For example, a child might hold up a ball (topic) and say "Wet" or "Blue" or "Mine" or "Roll" (comments). In predicating these things of the ball the child is clearly attempting to manipulate the attention (not just the behavior or gaze direction) of another person. She is asking the listener to attend to some specific aspect, out of other possible aspects, of their shared focus of attention—assuming that other persons can intentionally modulate their attention in response to linguistic and nonlinguistic means of communication, often while not changing their visual orientation at all.

The overall point is that in their imitative learning and declarative gesturing and, perhaps most of all, in their language learning and use human 1-year-olds demonstrate their understanding that adults have intentional control over their behavior and their attentional focus; that is, they understand them as intentional agents (Tomasello, 1995a; Tomasello, Kruger, & Ratner, 1993). These behaviors also demonstrate infants' understanding that people's focus of attention is underdetermined by the actual perceptual situation; that is, even though a child and an adult may be looking in the same direction, they may nevertheless have different foci of attention. Apes who have not been raised and trained in special ways by humans do not show evidence for anything like this kind of understanding of the intentionality of others.

What about Enculturated Apes?

In recent years, evidence has been mounting that apes raised and trained by humans in special ways—what my collaborators and I have called "enculturated apes" or, more neutrally, "human-raised apes"—are cognitively different in some ways from their wild and captive conspecifics. Call and Tomasello (in press) compiled data on ape cognition systematically as a function of their experience with humans and found that apes who have had extensive experience with humans consistently show more human-like skills of social cognition.

For example, apes raised by humans show more evidence of tuning into the intentional strategies of humans in imitative learning situations

(Tomasello, Savage-Rumbaugh, & Kruger, 1993), and they show more human-like skills of gesturing to humans and using human-like communicative symbols as well (Call & Tomasello, 1994; Savage-Rumbaugh, McDonald, Sevcik, Hopkins, & Rubert, 1986). Of most direct relevance to the current studies, (1) Gómez (1990) reported that a gorilla with much human experience often led caretakers by the hand to a desired but out-of-reach object, looking back and forth between the object and the human's eyes while doing so (see also Carpenter, Tomasello, & Savage-Rumbaugh, 1995); (2) Premack (1988) reported that a chimpanzee attempted to remove a blindfold from the eyes of a human whose help it needed to obtain food (although three other chimpanzees failed to do the same); and (3) Call and Tomasello (1994) found that one orangutan, who had been raised by humans and trained in some aspects of American Sign Language, manually indicated the food he wanted quite frequently when the human who could obtain it for him had his eyes open but did so only very infrequently when the human's eyes were closed (an orangutan who had a less rich set of experiences with humans did not differentiate between the eyes open and the eyes closed conditions). The suggestion is thus that interacting with humans, extensively and from early in ontogeny, may lead apes down more human-like developmental pathways in the domain of social cognition. (Perhaps Povinelli and Eddy's chimpanzees would have performed better if they had had more experience with humans.) This fact is powerful testimony to the power of human culture—including especially material and symbolic artifacts and the intentional instruction of others with regard to those artifacts—in shaping cognitive processes (Tomasello, 1995b). And it may even suggest indirectly that human cognitive development would not be what it is without the structuring role of culture as well.

Nevertheless, apes raised by humans do not thereby turn into humans. Although the limitations of human-raised apes' cognitive skills have never been systematically investigated, some differences with human children are still readily apparent. (1) Although human-raised apes request objects and activities from humans via various joint attentional strategies, they still do not regularly (if at all) simply show something to a human or an ape companion declaratively or point to something just for the sake of sharing attention to it (Gómez, Sarria, & Tamarit, 1993). (2) When compared with the skills of human children, ape skills with a human-like language system are limited in a number of ways (Tomasello, 1994a). (3) In tasks in which they must cooperate with conspecifics, without specific human training, ape skills of collaborative learning are curiously limited (Savage-Rumbaugh, Rumbaugh, & Boysen, 1978b). (4) Despite the reporting of some anecdotes (Fouts, Fouts, & Van Cantfort, 1989), there is very little, if any, behavior of human-raised apes that one would want to call intentional teaching.

The common factor in all these phenomena—that is, in declarative and linguistic communication, collaborative learning, and intentional teaching—is a sense of simply sharing experience with another psychological being, often for no reward other than the sharing itself. One hypothesis is thus that although apes can master the "referential triangle" in their interactions with humans for instrumental purposes when they are raised in human-like cultural environments they still do not attain a human-like social motivation for sharing experience with other intentional beings. This motivation to share may be what allowed human beings not just to tune into the culture of others but also to actually create culture themselves. We do not know precisely how or why human beings come to have such a motivation, but one possibility is that they identify their own psychological experiences with those of conspecifics in much deeper ways than do other animal species (Baressi & Moore, in press; Tomasello, 1995a).

Conclusion

Where does this leave us with regard to Povinelli and Eddy's important findings that chimpanzees know that others must be oriented toward them for their gestural signals to work but that they do not understand much about the role of the eyes in the process? One thing that these findings do not mean is that chimpanzees are not intelligent creatures; they are, and they express this intelligence in many and varied ways. The findings also do not mean that chimpanzees are necessarily incapable of understanding others as intentional beings; not appreciating the role of the eyes is not fatal to such an understanding. But in concert with the other studies and observations that I have reviewed here Povinelli and Eddy's elegant series of experiments contributes to a growing body of evidence suggesting that chimpanzees may not have a fully human-like understanding of the psychological lives of other intentional beings, be they humans or other chimpanzees.

If this emerging view of chimpanzees is correct—if chimpanzees and humans share many cognitive functions but humans also have a special social-cognitive adaptation—this has profound implications for our understanding of human evolution. Specifically, if it is true that a uniquely human form of social cognition both arose from and contributed to a cultural way of life, it is possible that this adaptation is of very recent origin. The material artifacts of prehistoric humans show basically no signs of cultural variability or cumulative historical change prior to the emergence of *Homo sapiens sapiens* less than 100,000–200,000 years ago (Klein, 1989)—several million years after humans and apes diverged from one another. This is also the first period during which we see material artifacts that are clearly symbolic or that seem to require collaborative or specialized efforts on the part of more than one person (Noble & Davidson, in press). If the cultural learning, symbolic communication, and collaborative labor that would seem to be necessary to produce these cultural artifacts may be taken as indicators of at least some form of intentional understanding (Tomasello et al., 1993), the implication is that the social-cognitive capacities on which human culture depends emerged only with modern human beings. The evidence from human evolution as manifest in material artifacts may thus be taken to support not just Povinelli and Eddy's "late" view of the emergence of human social cognition but an extremely late version of that late view.

References

Akhtar, N., & Tomasello, M. (in press). Intersubjectivity and early language. In S. Braaten (Ed.), *Intersubjective communication and emotion in ontogeny.* New York: Cambridge University Press.

Baron-Cohen, S. (1995). *Mindblindness: An essay on autism.* Cambridge, MA: MIT Press.

Barresi, J., & Moore, C. (in press). Intentional relations and social understanding. *Behavioral and Brain Sciences.*

Bernstein, I. (1988). Metaphor, cognitive belief, and science. *Behavioral and Brain Sciences,* **11,** 247–248.

Boesch, C. (1991). Teaching among wild chimpanzees. *Animal Behavior,* **41,** 530–532.

Boesch, C., & Boesch, H. (1989). Hunting behavior of wild chimpanzees in the Tai National Park. *American Journal of Physical Anthropology,* **78,** 547–573.

Call, J., & Tomasello, M. (1994). The production and comprehension of pointing in orangutans. *Journal of Comparative Psychology,* **108,** 307–317.

Call, J., & Tomasello, M. (1995). The use of social information in the problem-solving of orangutans and human children. *Journal of Comparative Psychology,* **109,** 308–320.

Call, J., & Tomasello, M. (in press). The role of humans in the cognitive development of apes. In A. Russon (Ed.), *Reaching into thought: The minds of the great apes.* Cambridge: Cambridge University Press.

Carpenter, M. (1995). *The social cognition of 9- to 15-month-old infants.* Unpublished doctoral dissertation, Emory University.

Carpenter, M., Tomasello, M., & Savage-Rumbaugh, E. S. (1995). Joint attention and imitative learning in children, chimpanzees and enculturated chimpanzees. *Social Development,* **4,** 217–237.

Chalmeau, R. (1994). Do chimpanzees cooperate in a learning task? *Primates,* **35,** 385–392.

Corkum, V., & Moore, C. (1995). Development of joint visual attention in infants. In C. Moore & P. Dunham (Eds.), *Joint attention: Its origins and role in development.* Hillsdale, NJ: Erlbaum.

Crawford, M. (1937). The cooperative solving of problems by young chimpanzees. *Comparative Psychology Monographs,* **14**(2).

de Waal, F. (1982). *Chimpanzee politics.* New York: Harper & Row.

de Waal, F. (1986). Deception in the natural communication of chimpanzees. In R. Mitchell & N. Thompson (Eds.), *Deception: Perspectives on human and nonhuman deceit.* Albany: State University of New York Press.

Fouts, R., Fouts, D., & Van Cantfort, T. (1989). The infant Loulis learns signs from cross-fostered chimpanzees. In R. Gardner, B. Gardner, & T. Van Cantfort (Eds.), *Teaching sign language to chimpanzees.* Albany: State University of New York Press.

Franco, F., & Butterworth, G. (in press). Pointing and social awareness: Declaring and requesting in the second year. *Journal of Child Language.*

Gómez, J. C. (1990). The emergence of intentional communication in the gorilla. In S. Parker & K. Gibson (Eds.), *"Language" and intelligence in monkeys and apes.* Cambridge: Cambridge University Press.

Gómez, J. C., Sarria, E., & Tamarit, J. (1993). The comparative study of early communication and theories of mind: Ontogeny, phylogeny, and pathology. In S. Baron-Cohen, H. Tager-Flusberg, & D. J. Cohen (Eds.), *Understanding other minds: Perspectives from autism.* New York: Oxford University Press.

Harcourt, S., & de Waal, F. (1992). *Coalitions and alliances in humans and other animals.* Oxford: Oxford University Press.

Heyes, C. (1993). Anecdotes, training, trapping, and triangulating: Can animals attribute mental states? *Animal Behavior,* **46,** 177–188.

Klein, R. (1989). *The human career.* Chicago: University of Chicago Press.

Kruger, A., & Tomasello, M. (in press). Cultural learning and learning culture. In D. Olson (Ed.), *Handbook of education and human development.* Oxford: Blackwell.

Meltzoff, A. (1988). Infant imitation after a one week delay: Long term memory for novel acts and multiple stimuli. *Developmental Psychology,* **24,** 470–476.

Meltzoff, A. (1995). Understanding the intentions of others: Re-enactment of intended acts by 18-month-old children. *Developmental Psychology,* **31,** 838–850.

Menzel, E. (1971). Communication about the environment in a group of young chimpanzees. *Folia Primatologica,* **15,** 220–232.

Nagell, K., Olguin, K., & Tomasello, M. (1993). Processes of social learning in the tool use of chimpanzees and human children. *Journal of Comparative Psychology,* **107,** 174–186.

Noble, W., & Davidson, I. (in press). *Human evolution, language, and mind: A psychological and archaeological inquiry.* Cambridge: Cambridge University Press.

Povinelli, D. (1994). Comparative studies of animal mental state attribution: A reply to Heyes. *Animal Behaviour,* **48,** 239–241.

Povinelli, D., Nelson, K., & Boysen, S. (1990). Inferences about guessing and knowing by chimpanzees. *Journal of Comparative Psychology,* **104,** 203–210.

Premack, D. (1986). *Gavagai!* Cambridge, MA: MIT Press.

Premack, D. (1988). "Does the chimpanzee have a theory of mind?" revisted. In R. Byrne & A. Whiten (Eds.), *Machiavellian intelligence: Social expertise and the evolution of intellect in monkeys, apes, and humans.* New York: Oxford University Press.

Premack, D., & Woodruff, G. (1978). Does the chimpanzee have a theory of mind? *Behavioral and Brain Sciences,* **4,** 515–526.

Savage-Rumbaugh, S., McDonald, K., Sevcik, R., Hopkins, W., & Rubert, E. (1986). Spontaneous symbol acquisition and communicative use by pygmy chimpanzees (*Pan paniscus*). *Journal of Experimental Psychology: General,* **115**(3), 211–235.

Savage-Rumbaugh, S., Rumbaugh, D., & Boysen, S. (1978a). Commentary on Premack and Woodruff (1978). *Behavioral and Brain Sciences,* **4,** 527–529.

Savage-Rumbaugh, S., Rumbaugh, D., & Boysen, S. (1978b). Symbolic communication between two chimpanzees (*Pan troglodytes*). *Behavioral and Brain Sciences,* **1,** 539–554.

Tomasello, M. (1990). Cultural transmission in the tool use and communicatory signaling of chimpanzees? In S. Parker & K. Gibson (Eds.), *"Language" and intelligence in monkeys and apes: Developmental perspectives.* Cambridge: Cambridge University Press.

Tomasello, M. (1994a). Can an ape understand a sentence? A review of *Language comprehension in ape and child* by Savage-Rumbaugh et al. *Language and Communication,* **14,** 377–390.

Tomasello, M. (1994b). The question of chimpanzee culture. In R. Wrangham, W. McGrew, F. de Waal, & P. Heltne (Eds.), *Chimpanzee cultures.* Cambridge, MA: Harvard University Press.

Tomasello, M. (1995a). Joint attention as social cognition. In C. Moore & P. Dunham (Eds.), *Joint attention: Its origins and role in development.* Hillsdale, NJ: Erlbaum.

Tomasello, M. (1995b). The power of culture: Evidence from apes. *Human Development,* **38,** 46–52.

Tomasello, M. (in press-a). Do apes ape? In J. Galef & C. Heyes (Eds.), *Social learning in animals: The roots of culture.* New York: Academic.

Tomasello, M. (in press-b). Perceiving intentions and learning words in the second year of life. In M. Bowerman & S. Levinson (Eds.), *Conceptual development and the acquisition of language.* Cambridge: Cambridge University Press.

Tomasello, M. (in press-c). Social cognition and the evolution of culture. In J. Langer & M. Killen (Eds.), *Piaget, evolution, and development.* Hillsdale, NJ: Erlbaum.

Tomasello, M., & Barton, M. (1994). Learning words in non-ostensive contexts. *Developmental Psychology,* **30,** 639–650.

Tomasello, M., & Call, J. (1994). The social cognition of monkeys and apes. *Yearbook of Physical Anthropology,* **37,** 273–305.

Tomasello, M., Call, J., Nagell, K., Olguin, K., & Carpenter, M. (1994). The learning and use of gestural signals by young chimpanzees: A trans-generational study. *Primates,* **35,** 137–154.

Tomasello, M., Davis-Dasilva, M., Camak, L., & Bard, K. (1987). Observational learning of tool-use by young chimpanzees. *Human Evolution,* **2,** 175–183.

Tomasello, M., Gust, D., & Frost, T. (1989). A longitudinal investigation of gestural communication in young chimpanzees. *Primates,* **30,** 35–50.

Tomasello, M., Kruger, A., & Ratner, H. (1993). Cultural learning. *Behavioral and Brain Sciences,* **16,** 495–511.

Tomasello, M., Savage-Rumbaugh, S., & Kruger, A. (1993). Imitative learning of actions on objects by chimpanzees, enculturated chimpanzees, and human children. *Child Development,* **64,** 1688–1705.

Tomasello, M., Strosberg, R., & Akhtar, N. (in press). Eighteen-month-old children learn words in non-ostensive contexts. *Journal of Child Language.*

Whiten, A., & Byrne, R. (1988). Tactical deception in primates. *Behavioral and Brain Sciences,* **11,** 233–244.

Woodruff, G., & Premack, D. (1979). Intentional communication in the chimpanzee: The development of deception. *Cognition,* **7,** 333–362.

REPLY

GROWING UP APE

Daniel J. Povinelli

Over 2 years have now elapsed since Timothy Eddy and I completed the series of experiments reported in this *Monograph*. Meanwhile, our seven apes have grown 2 years older. This passage of time has allowed us to conduct some additional research on these same animals that has direct implications for the important theoretical points raised by both Hobson and Tomasello as well as for some of the hypotheses we entertained at the end of Chapter VI. In particular, longitudinal follow-ups using some of the same tasks have provided us with some intriguing new information that further constrains the interpretations of our data set. In this Reply, I draw on this new work to address two recurring and interrelated themes in both our *Monograph* and the Commentaries: (1) the relation between experience and conceptual understanding in theory of mind development and (2) the relation between performances based on behavioral/procedural rules and those that are augmented by an additional, conceptual understanding of mental states. Finally, I use this discussion to sketch an alternative view that could account for the confusing mosaic of similarities and differences we see when comparing the behavioral and cognitive systems of humans and apes.

Is Experience the Master?

Amount of Experience

Anyone reading this *Monograph* will note our hesitation in automatically generalizing our findings to adult chimpanzees—a concern that Tomasello underscores in his Commentary. Indeed, one of the alternative hypotheses that we outlined in Chapter VI was that although 5–6-year-old chimpanzees do not seem to understand seeing-as-attention, older chimpanzees might. To put it crudely, apes grow up, too. Readers will also note our insistence that although the pattern of our subjects' performance was best predicted by a learning framework, the very fact that in the end they were quite adept at choosing the experimenter who could see them—no matter how we posed the question—left us in a difficult quandary. On the one hand, it was possible (indeed likely) that with experience our apes had learned a set of procedural rules that were sufficient to solve our tasks, without any concomitant understanding of attention or "seeing" per se. On the other hand, even though the emergence of these rules can be easily explained in terms of general learning theory, it is also possible that this experience simultaneously endowed our apes with a genuine understanding of seeing-as-attention. As we noted, choosing between these two alternatives is not easy. More important, either choice has implications for our interpretation of similar research with young children. For example, if our apes came to provide the appearance (but not the reality) of a mentalistic understanding of seeing, who is to say that the children we tested were not in the same position? To be sure, the degree to which apes or children are able to generalize to new situations may give us some indication of what they have really learned, but a moment's reflection will reveal that there are ways in which such generalization could occur without an attendant conceptual theory of seeing.

Tomasello misinterprets our raising this warning flag as meaning that we believe that very young children may not really possess theory of mind skills. To the contrary, I have been convinced for quite some time by the same body of evidence to which Tomasello refers (some of which he has been instrumental in collecting) that by 18–24 months human toddlers have taken their first firm steps into a psychological arena in which the social milieu is understood in terms of private mental states (for an early statement of this position, see Gallup & Suarez, 1986). But understanding the conglomerate of changes that occur at this general developmental period (18–24 months) can tell us only so much about the underlying cognitive processes supporting any given behavior that emerges during (or after) this period. As a case in point, many commentators (including Tomasello) have expressed doubt about the cognitive significance of the ability of many chimpanzees to recognize themselves in mirrors. These criticisms are worth consideration, but the ability to recognize oneself in mirrors typically emerges in human infants at about 18–24 months (Amsterdam, 1972; Lewis & Brooks-Gunn, 1979). So in this case the fact that a skill emerges alongside a cluster of others is not a necessary indication of its underlying psychological cause (for data concerning the correlation between the emergence of self-recognition in mirrors and other behaviors, see Asendorpf & Baudonniere, 1993; Bischof-Köhler, 1988, 1994; Johnson, 1982; Lewis, Sullivan,

Stanger, & Weiss, 1989). Likewise, at an even more detailed level of analysis, Gopnik and Meltzoff (1986) have provided theoretical and empirical grounds for suspecting that many of the cognitive achievements during this period may be occurring in relatively separate domains.

Thus, the fact that young 2-year-olds have some understanding of mental states such as desire and attention, but that 2½-year-olds did poorly on some versions of our test of seeing-as-attention, does not uniquely call the validity of our task into question. Furthermore, it says little one way or the other about how competence on such tasks is constructed in the first place. Clearly, our question remains as salient now as it was 2 years ago: How do we distinguish between competence on theory of mind tasks that is the result of procedural rules derived from sheer experience versus competence that is both derived from experience and accompanied by conceptual understanding? Transfer tests of the type that we have employed here and elsewhere can help, but the manner in which we have constructed the problem may leave one wondering if we are left forever chasing a moving target.

Fortunately, one way of distinguishing between these alternatives serendipitously presented itself to us 13 months after the completion of the final study with our apes that we reported in this *Monograph*. In the context of a very different set of studies, we once again administered several of the seeing-versus-not-seeing treatments to our subjects. Initially, our protocol merely called for administering the eyes-open-versus-closed (C'''') treatment to show that our apes were still sensitive to this distinction and would respond in the same fashion as they had in the studies reported in this *Monograph*. Much to our surprise, however, the animals no longer seemed to have a preference for the experimenter who had his or her eyes open (Fig. R1). Even after 48 blocked trials of this treatment using the same stimulus configurations outlined in Experiment 13 (Chap. IV), our seven apes were responding at chance levels. Puzzled by this finding, we turned to the screens-versus-no-screens (B'') treatment because we believed that it was more visually salient and because our apes had had considerably greater experience with it than with the eyes-open-versus-closed treatment. Again, however, the animals performed randomly—only gradually showing evidence of learning across the 12 probe trials they received (Fig. R1). Stepping back, we decided to administer four probe trials of the back-versus-front (C) treatment. After all, this was a treatment that the subjects had "understood" immediately nearly 2 years earlier. And, just as we expected, the animals' performance shot up to near perfect levels (Fig. R1).

The results of this longitudinal follow-up can be interpreted in a number of ways, but in the interest of space let me focus on just two. First, a strict learning theorist has a ready-made explanation of these findings: in the course of participating in a large number of additional studies in the intervening 13 months, our subjects had long since abandoned or forgotten

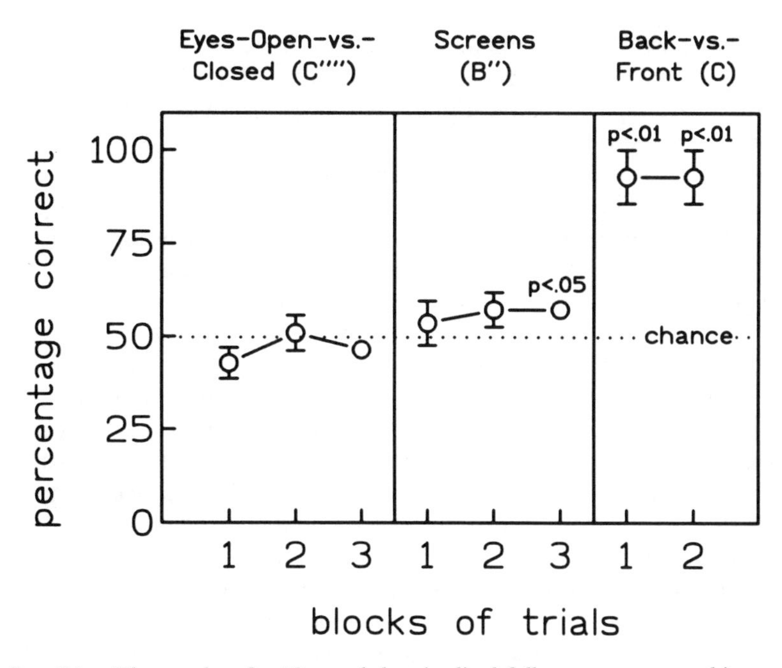

FIG. R1.—The results of a 13-month longitudinal follow-up on seven chimpanzees' performance on three of the experimental treatments used in Experiments 1–14. The subjects' mean percentage correct (± SEM) is shown in blocks of trials (C'''' in blocks of 16 trials, B'' in blocks of four trials, C in blocks of two trials). Trials were administered using the same general methods as described in Chaps. III and IV.

their old rules, having updated, revised, and/or replaced them numerous times. In contrast, their immediate and excellent performance on the back-versus-front (C) trials indicates their retention of a very strong rule or social disposition—a rule or disposition that continued to be useful in many of the additional studies in which they had participated. (For example, although only one experimenter had been present in most of these intervening studies, the apes were nonetheless required to orient to this person. Therefore, being attracted in general to the frontal stimulus of a human was still a very useful behavioral rule.)

I am uneasy about arguing too strongly in favor of this narrowly focused learning interpretation because its implications are so far-reaching. For example, if this explanation is correct, it means that our apes' generally successful performance at the end of the studies reported in Chapters III and IV can be clearly and conceptually distinguished from the successful performance of the young children we tested. Even if our young children had received numerous semistructured experiences with instances of "seeing" and "not seeing" prior to participating in our tests (so many in fact that their first trials on our task might be more equivalent to the apes' final trials), they may have nonetheless parted conceptual company with the apes early on. For example, imagine these same children returning to our laboratory a year after participating in our tests and being retested on some new and arbitrary test of seeing-as-attention. The outcome cannot be in doubt. In the course of their development, through some combination of experience and hardwiring, they have constructed an enduring, fundamental understanding of the social universe—a system of understanding other minds (and maybe even the world as a whole) that transcends specific rules needed to navigate fairly narrow regions of ecological space, a system of understanding that we might truly wish to adorn with the label *theory* (see Gopnik & Meltzhoff, in press). But not, apparently, our apes. Although they are quite clever, our chimpanzees' striking difficulty on their longitudinal follow-ups provides learning theorists with additional support for their interpretations. The apes appear tied to social dispositions and procedural rules that are forged by epigenetic interactions, executed with precision, generalized to similar situations, but never integrated as part of a broad, interpretive apparatus with which to understand the minds of those around them.

Yet the learning theorist is not alone in vying for our conceptual allegiance; other interpretations of these longitudinal data are available. For example, we might imagine that when it comes to the development of theory of mind, experience is less the master than Alison Gopnik and others have speculated (e.g., Gopnik, 1993; Gopnik & Meltzoff, in press). Instead, the epigenetic pathways controlling the expression of theory of mind abilities in humans (and possibly in other species as well) may be tightly canalized. Across a fairly wide range of environments and experiences, the exact timing of the development of their cognitive capacities may be fairly uniform. The timing of the apparent conceptual transitions detected during infancy and young childhood may be real and may reflect maturational changes in neural systems subserving theory of mind skills.

However, a thorough knowledge of the rate and timing of human development in theory of mind will not necessarily guarantee an accurate understanding of the rate and timing of chimpanzee development. Indeed, there is some evidence that certain general aspects of cognitive development are temporally compressed in humans, relative to chimpanzees (see Povinelli, 1994, 1996). From this perspective, the longitudinal data reported in Figure R1 could simply mean that we should extend upward to about 7 years the lower age limit at which apes might develop a theory of seeing-as-attention. In our present state of ignorance, we should seriously entertain this possibility. But if we continue to obtain similar results as those reported here as these apes' adulthood approaches (and as we begin to work directly with adult apes), at some point we will be forced to abandon this developmental explanation of their differences from young children. Indeed, as most of our apes' eighth birthdays are approaching, they continue to respond to even our simplest tests of an appreciation of attention in ways best predicted by a learning framework (see Povinelli, Bierschwale, Reaux, & Čech, 1996).

Kinds of Experience

If the sheer amount of experience on a series of tests may or may not alter underlying comprehension, what about the *kinds* of experience that a chimpanzee might have? For example, what if one were to grow up an ape, but in the company and culture of humans? Could this experience significantly alter the epigenetic course and outcome of a chimpanzee's cognitive development? Tomasello turns our attention to "enculturated" apes as a case in point and wonders whether more experience with humans might not have significantly affected our apes' performances. Some of his own research on imitation suggests that chimpanzees raised with humans and adorned with their material culture develop at least a rudimentary understanding of the intentions of others (Tomasello, Savage-Rumbaugh, & Kruger, 1993).

To begin, I do not possess any strong a priori feelings about whether apes reared by humans undergo profound psychological reorganizations. However, I do not think that the strength of the empirical evidence to date forces us to take the view that an understanding of intentions—which on Tomasello's reading normally eludes the chimpanzee's mental system—can be imbued by raising them with another species whose cognitive/behavioral system is so constructed. The best data sets available with which to evaluate his hypothesis concern the capacity for imitation, which his studies suggest may not be present in wild chimpanzees but may emerge in chimpanzees enculturated with humans (Tomasello et al., 1993). However, there is at least some data that indicate that imitation may be present in chimpanzees no more enculturated (and some certainly less so) than our subjects (see Custance, Whiten, & Bard, in press; Whiten, Custance, Gómez, Teixidor, & Bard, 1996). Thus, it may be difficult to argue that enculturation is responsible for the *de novo* appearance of true imitative abilities.

In other domains such as chimpanzees' understanding of seeing, the evidence that Tomasello marshals in support of the enculturation hypothesis is often circumstantial, limited to quasi-experimental approaches that have been only incompletely reported (i.e., Premack's, 1988, account of how one of four chimpanzees attempted to remove an experimenter's blindfold before enlisting his assistance on a task). In addition, because the relevant developmental inputs are not precisely specified (although for Tomasello they are related to pragmatics more that language per se), it is difficult to know how to apply the enculturation hypothesis uniformly. For instance,

Tomasello rejects Premack and Woodruff's (1978b) landmark study of a highly enculturated adult chimpanzee's understanding of intention by arguing that the task (selecting a photograph that fulfills the goal of an actor on videotape struggling to solve a staged problem) could be completed by simple associative rules. However, he does not mention Premack and Woodruff's (1978a) theoretical refutation of this objection or Premack and Dasser's (1991) empirical data with young children demonstrating the implausibility of this counterinterpretation. My point is neither that Premack and Woodruff's (1978b) tests with Sarah definitely established that chimpanzees understand intention nor that Tomasello's enculturation hypothesis is incorrect. Rather, it should be clear that accounts of this kind are a double-edged sword for the enculturation hypothesis—if Sarah merely used clever association-based algorithms to solve this task, then a similar kind of analysis could be applied to other enculturated apes on other tasks as well.

Having expressed my reservations about the enculturation hypothesis, let me now explore some of the implications of accepting (or rejecting) it. First, if mere exposure to the conventions and material culture of humans can transform their understanding of others so profoundly, then we might no longer wish to entertain the idea of dedicated cognitive modules that evolved explicitly for the purpose of generating inferences about the mental states of others (e.g., Baron-Cohen, 1995; Fodor, 1992; Leslie, 1994). If chimpanzees (and other nonhuman primates) normally do not possess the cognitive abilities that modularity theorists envision as being controlled by specialized, evolved brain modules, but if enculturated chimpanzees do, then obviously the underlying neural systems subserving theory of mind in humans did not originally evolve for that purpose. If true, our neurobiological account of theory of mind would need to be recast in terms of the evolution of developmental systems that retain a high degree of plasticity in the face of different epigenetic environments.

From the opposite point of view, however, let us make the assumption that the typical absence of the expression of theory of mind in chimpanzees can be generalized to older animals and furthermore reflects the unique evolution in humans of some kind of neural system controlling its expression. That is, let us imagine that cognitive specializations in theory of mind (regardless of its exact neural basis) evolved in humans after our lineage split from the line leading to the African apes (Povinelli & Preuss, 1995). How, then, could we account for the apparent cultivation of theory of mind in enculturated apes? Again, one possibility is that its presence is more apparent than real. Although the behavioral end product may be similar, enculturated apes may have arrived via a different route. One of the clear lessons from this *Monograph* is that with sufficient experience (perhaps of exactly the type received by enculturated animals) chimpanzees may act as if they understood the intentionality of seeing, for example, without really doing so. In this case, the "curious limitations" that Tomasello notes in the enculturated ape's social understanding may not be limitations at all. Rather, they may merely reflect the holes left in the nonmentalistic cognitive scaffolds constructed by one species forced to cope with a different species that expects such social understanding.

Finally, a limitation of the enculturation hypothesis is that it never clearly specifies which inputs are critical or how these inputs inbue the ape's cognitive system with an understanding of the mental life of others. This is not a criticism unique to the enculturation hypothesis. Rather, it exposes broad, conceptual uncertainties surrounding our ideas of how theory of mind abilities are constructed and develop during the course of human ontogeny. This aside, consider Tomasello's closing comments that the evolution of human culture may be dependent on the evolution of skills related to social cognition. I agree, and we have presented a detailed argument about how these cognitive skills explain both the workings and evolution of universal aspects human ethics and morality and the absence of such systems in chimpanzees (Povinelli & Godfrey, 1993). Yet this agreement masks some important unresolved issues concerning both the nature of the adaptations that produced these systems in humans in the first place and the epigenetic environments that trigger their development now. One possibility is that this epigenetic system reflects underlying changes in the genes controlling development of specific neural systems that were selected for during the course of human evolution. But if this is true, how can mere enculturation with humans trigger its expression in a species that descended from ancestors who (by definition) split off from the line leading to humans prior to the evolution of this adaptation? One possible answer is that forcing the ape's developing system to accommodate the content and practice of certain aspects of human culture may somewhat alter its typical ontogenetic course. The theoretical challenge for the enculturation hypothesis is, How much and how fundamentally can this course be altered?

What about Language?

Thus far, the causal role of human language has played a small part in this discussion of the enculturation hypothesis. Yet it is quite possible that some aspects of training apes in regimes designed to have them produce and comprehend the semantic and syntactic features of human language may result in deep, foundational changes in their conceptual systems. For example, Premack (1988) argued that training apes to use a symbol for same/different judgments about objects produced a cognitive system that could understand the relations critical to solving analogies. Is it similarly possible that exposure to certain aspects of a human language system (pragmatic, semantic, or syntactic) may directly alter the ape's theory of mind? This question raises a variant of the even broader classic question concerning the relation between language and mind: Is theory of mind possible in the absence of a communication system with the semantic and syntactic features of human language?

Assessing the effect of language training on a chimpanzee's theory of mind raises a number of extremely difficult challenges. First, the strong version of the language transformation hypothesis hinges on whether an ape's exposure to and use of aspects of a human language system forces it to comprehend the various aspects of that language system (pragmatic, syntactic, semantic) in the way humans do. Second, there is the question of whether mental states (as concepts) are or can be represented in the absence of a linguistic code that can compress as much information as human language. This, of course, is the classic mind-language problem. It is possible that apes (and other nonhuman animals) have mental codes that allow for the use of images to represent something like epistemic and nonepistemic mental states. Indeed, we could even imagine that apes possess a syntactic device for parsing and then productively reordering such images in order to generate novel strings of images (ideas). (On the other hand, I have often wondered whether many of the differences between apes and humans are the consequence of a very limited ability of apes to engage in such syntactic construction and deconstruction of the behavior stream. That is, perhaps humans have specialized in an ability to decompose a given segment of the behavior stream in almost infinitely unique ways. Certainly, the sophistication of this ability would have direct implications for the extent to which the behavior of apes is exclusively governed by a complex interaction of social dispositions and learned procedural rules.)

Yet even if wild chimpanzees possess a system for manipulating an image-based code—a system powerful enough to produce a sufficient distance of sign from referent that we would wish to gloss some given image as a representation of a "mental state"—we still must acknowledge the fact that if language-trained apes come to use symbols to compress information on the order of magnitude that humans do, then perhaps they may find themselves in a position to distance the observed actions of others from idealized instances of those actions. In short, if language-trained apes come to use symbols as a shorthand for compressing many ideas about objects or events into a single unit, this could provide a similar vehicle for generating a kind of shorthand, heuristic code for representing the potential actions of others—a code perhaps not so different from the foundational aspects of our own folk psychology. If this were the case, then the limitations on their ability to understand mental states would be related to the extent to which their inherited representational systems can accommodate compression of action potentials into "ideas" about such unobservable states. Furthermore, it would not be exposure to the pragmatics of human language that would facilitate an understanding of mental states but rather the semantic and syntactic features of language. Of course, all this hinges on our ability to resolve the first point raised above. That is, if the behavioral productions of these abnormally reared, language-trained apes are the result of learned rules and procedures unrelated to the syntactic and semantic features of human language, we would have no reason to entertain these ideas (for empirical data bearing on this question, see Savage-Rumbaugh et al., 1993).

Are Apes ''Mere Behaviorists''?

In this Reply, I have continued to maintain our position that our apes' behavior on our tests is best predicted by a learning framework. However, as is his gift, Hobson grapples with the details of this claim and exposes what is, in truth, a far more complicated portrait than we have outlined. He is correct in assuming from our disclaimers that we are in full agreement that a spectrum of possibilities exists concerning young chimpanzees' and young humans' understanding of the subjective aspects of any given facet of mental life. One's understanding of the social world is not necessarily either behaviorist or mentalistic. Elements of both can and do coexist alongside each other in humans, and they may in other species as well.

In addition, to be a mentalist one may have either a quite elaborate understanding of mental states and events (as do adult humans) or but a fledgling appreciation of the subjective aspect of self and other (as perhaps do human infants and children). Similarly, Tomasello underscores our point that the apes' understanding of seeing need not be as sophisticated as young 3-year-olds' in order to qualify as genuinely mentalistic (after all, even our 2½-year-olds did not always do well). Indeed, as we noted in our second alternative hypothesis, there may be ways in which apes (or human infants and children) can understand attention other than understanding the eyes or face as portals through which attention emanates. Furthermore, as Hobson points out, chimpanzees or human infants may also react to others in ways that suggest that they register (or directly perceive) the subjectivity of another's expressive attitude or feeling, without being able to represent that this is to what they are responding. Indeed, in posing these challenges, Hobson has anticipated that much of our current work (despite our earlier protestations of the theoretical difficulties involved) bears directly on both these questions (Povinelli et al., 1996; Povinelli & Eddy, 1996, in press).

There are several aspects of Hobson's insistence that we consider all points on the "spectrum of possibilities" that I embrace wholeheartedly.

First, as he points out, we need an account of how infants develop an appreciation of the subjectivity of self, "selves," and the mental world in general that is not discontinuous—an explanation that does not leave the infant suddenly leaping from behaviorist to mentalist. Important in this kind of explanation will be the kinds of psychological configurations that an infant's developing brain can sustain short of a full appreciation of the content and character of adult mental states, attitudes, and events. Although in places I have trouble understanding why some of his "attitudes" could not be described in strictly procedural terms, in general the point is well taken. Second, I agree that we need an explanation of how and when evolution has acted to produce this kind of ontogenetic system in the first place. Indeed, John Cant and I have recently attempted to provide an evolutionary account of at least one strand of this puzzle, the evolutionary emergence of understanding the self as a causal agent (Povinelli & Cant, 1995).

The qualitative versus quantitative change problem that Hobson grapples with is an old one in comparative psychology. Indeed, it is so old that many comparative psychologists dismiss it as archaic, a nonissue. I disagree. As we pointed out in Chapter I, this problem remains a conceptual challenge to developing a comparative psychology of mind. Yet in expressing my appreciation of the difficulty of these issues I must also express my concern about how one might interpret Hobson's comments. One line of reasoning typically goes like this: (1) humans possess a subjective understanding of self and other; (2) these abilities evolved (they were not divinely given); (3) chimpanzees are demonstrably very closely related to humans; and, therefore, (4) chimpanzees must possess at least rudiments of the same qualitative kind of subjective understanding of self and other that humans develop. As I have pointed out elsewhere, this inference does not follow (see Povinelli, 1993). When it comes to understanding the continuous chain of processes that have led to the evolution of a given biological system or structure, it is one thing to speak of precursors, foundations, or building blocks, but quite another to tackle the more difficult problem of function. Because we inhabit only a very narrow slice of evolutionary time, we are privy to only a thin cross section of the diversity that it has produced—a cross section that, because of extinction and genetic/morphological/behavioral evolution within populations, does not retain all the continuous branches of the tree we wish to reconstruct (see Fig. R2). Extinction, coupled with the imperfections of the fossil record, virtually guarantees that during periods of rapid evolutionary change we can never accurately identify the exact origin and subsequent elaboration of a given trait. Simply put, when it comes to evolution, discontinuous patterns do not necessarily imply discontinuous processes.

The point I wish to make most strongly is this: chimpanzees and other great apes and nonhuman primates may or may not share with humans

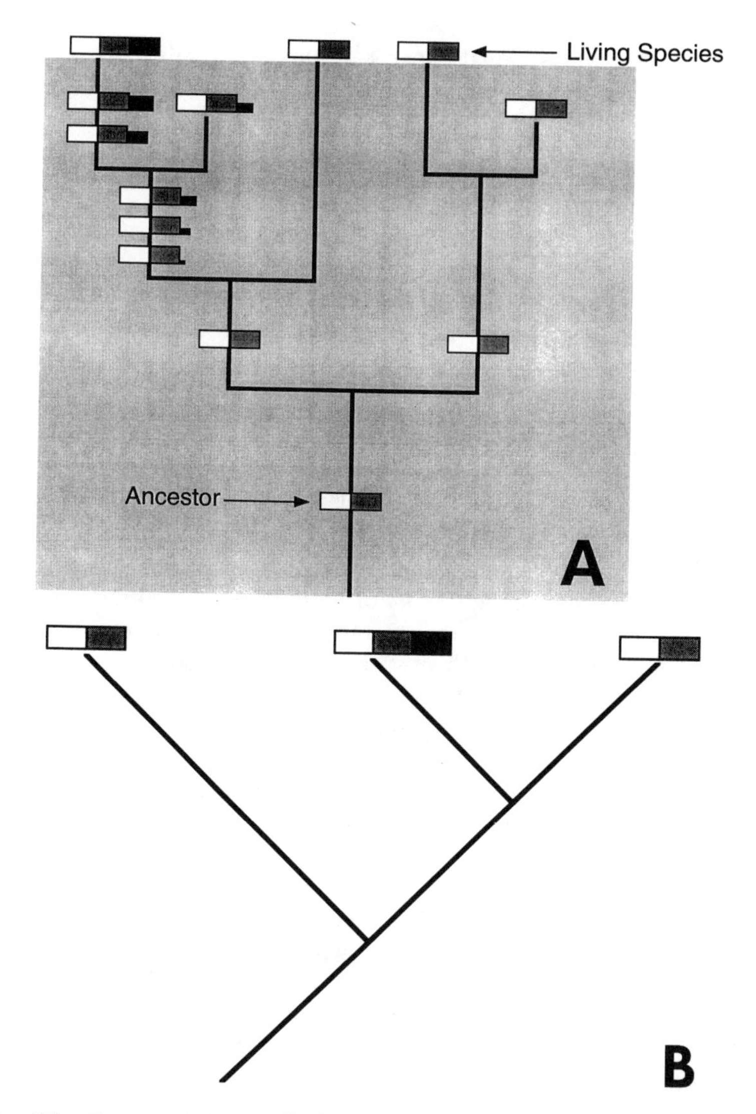

FIG. R2.—In *a,* a modern radiation of some arbitrary clade of species is shown, complete with all ancestors and descendants. In addition, the evolutionary emergence of a hypothetical cognitive-behavioral system (dark box) is tracked through evolutionary time from its initial appearance to its later elaboration. For purposes of clarity, this elaboration is depicted as the trait increasing in size. Extinct species inhabit the shaded region, whereas the three living species appear in the unshaded area. In *b,* just the living representatives of the radiation are shown as a part of a modern clade. Note that in *a* there is no dramatic step from the absence of the new system to its presence (although even here its initial appearance may have occurred as a result of a qualitative shift in function from previous, related systems). However, in *b* the appearance of the novel system appears discontinuous because of the fact that living species almost never comprise true ancestor-descendant relationships and because of the imperfections of the fossil record. Indeed, in the case of the evolution of cognitive systems, it is not clear whether even a pristine fossil record could yield sufficient morphological information (i.e., brain endocasts) to infer the presence or absence of the cognitive system.

limited, perhaps foundational aspects of theory of mind. However, our preconceived ideas about the rate and timing of evolutionary change cannot produce an a priori answer to the question. In this respect, evolutionary reconstructions are part of a historical science that is impenetrable to intuitions about how things ought to have evolved.

An Alternative View

In the final section of this Reply, I wish briefly to sketch an alternative view to the fairly simplistic notion of the evolution of cognitive development that has permeated my own thinking and research (as well as, I believe, that of others) for a number of years. I hope to show how the confusing array of similarities and differences that we see between humans and chimpanzees may be understood within the framework of modern biological ideas about the evolution of development (i.e., heterochrony).

Imagine either that humans have exclusively evolved a specialization in theory of mind or that the ancestors of modern humans inherited a fairly limited mentalistic psychology that was then radically elaborated and transformed during the course of their separate 5–7-million-year radiation as a separate lineage of bipedal apes. In either of these cases, humans can be envisioned as possessing an elaborated psychological system for generating, sustaining, and revising representations of the intentions, goals, desires, beliefs, thoughts, and activities of themselves and others. If this is true, what are we to make of spontaneously occurring behaviors (such as desception) that humans and apes share in common but that humans alone appear to understand in a thoroughly mentalistic manner? Take the case of gaze following (see Experiment 12, Chap. IV; Povinelli & Eddy, 1996). Both humans and chimpanzees have been shown to be capable of elaborate forms of this ability. Some researchers see the emergence of this behavior as signaling the arrival of the capacity for joint or shared attention in human infants (e.g., Baron-Cohen, 1995). However, one skeptical possibility is that this behavior has nothing whatsoever to do with the infant's understanding of the attentional focus of others. Rather, it could be that a later-developing psychological system in humans (their theory of mind) allows for a retrospective interpretation of gaze in this manner—a reinterpretation that is never allowed for by the ape's psychological system.

However, it is possible that we need to be willing to think differently about certain behaviors depending on whether it is an ape or a human infant that is displaying them. It could be that when apes engage in so-called shared attention behaviors, they do so without a co-occurring assignment of meaning, but that when human infants engage in the very same behaviors, such an assignment is made. Before I am criticized for using a double standard, it would be worthwhile to entertain the possibility that gaze following, deception, and the like are fairly ancient *behavioral* mechanisms that evolved long before anything like theory of mind had evolved. Maybe it is our species alone that has evolved a capacity to represent each other's mental states and in doing so has been placed in a position of being able to reinterpret behavioral patterns that evolved long before we did. Consider the following. After following someone else's gaze, when is it that we wonder what it is that she is looking at? Is the gaze-following response triggered by an inference about the mental state, or does exactly the opposite causal relation hold? Or perhaps it is something intermediate between the two. We have argued that much of the utility of theory of mind occurs in exactly this retrospective manner—as a mechanism for planning future actions, not as a means of solving immediate social problems (Povinelli & Povinelli, 1996). In short, where some researchers tend to describe social primates as having evolved a coherent theory of mind to cope with on-line social problems, we envision nearly the opposite. What social mammals need most is to act quickly, and what better for this than the ability to form a powerful set of automatized, procedural social rules from which they can quickly select an appropriate series of behavioral scripts. In this view, theory of mind is merely a reflective, after-the-fact, added bonus.

Thus, in our species (at least) a connection appears to exist between expressions of certain behaviors and a retrospective mentalistic interpretation of those behaviors. This may reflect the operation of cognitive innovations peculiar to our species that in the course of human ontogeny now emerge alongside more ancestral behaviors that evolved in isolation from this system. There are really at least two possible versions of this idea. First, the innovations in cognitive development may express themselves very early in the developing human infant, so that during each phase of cognitive development a psychological mechanism related to theory of mind allows for a qualitatively different kind of interpretation of behavior than what is present in other species (Povinelli, Zebouni, & Prince, 1996). A second possibility is that the behaviors that some researchers like Hobson interpret as evidence of a kind of intersubjectivity or shared affect or attention are really no more than behavioral algorithms that emerge through epigenetic interactions during the course of infant development. As opposed to tinkering with the earliest phases of cognitive development, selection may have modified later phases to gradually produce an epigenetic system that could support representations of mental states. For example, it is quite compatible with a skeptical view of the developmental literature in this area that human infants must await their second year of life before they develop the evolutionarily novel cognitive structures that allow them to see their own behavior and that of others in genuinely mentalistic terms.

Although I believe that it is a hypothesis worthy of consideration, I cannot demonstrate with any degree of certainty that the alternative view that I have sketched here is correct. Neither can Tomasello demonstrate unequivocally that enculturated apes develop an interpretive framework that shares many of the features of our own theory of mind system. Likewise, Hobson cannot establish without doubt that 1-year-old human infants are engaged in a genuinely mentalistic attitude toward others or that chimpanzees share this stance. Nor, indeed, do I think any of us would want to make such strong claims in the face of so young a science. Rather, their Commentaries and this Reply follow a more rewarding (and ultimately more productive) path of using current results to understand the shape of the remaining theoretical possibilities.

References

Amsterdam, B. (1972). Mirror self-image reactions before age two. *Developmental Psychobiology,* **5,** 297–305.

Asendorpf, J. B., & Baudonniere, P.-M. (1993). Self-awareness and other-awareness: Mirror self-recognition and synchronic imitation among unfamiliar peers. *Developmental Psychology,* **29,** 88–95.

Baron-Cohen, S. (1995). *Mindblindness: An essay on autism.* Cambridge, MA: MIT Press.

Bischof-Köhler, D. (1988). Uber der Zusammenhang von Empathie und der Fahigkeit, sich im Spiegel zu erkennen [On the association between empathy and ability to recognize oneself in the mirror]. *Schhweizerische Zeitschrift für Psychologie,* **47,** 147–159.

Bischof-Köhler, D. (1994). Selbstobjektivierung und fremdbezogene emotionen [Self-objectification and other-oriented emotions]. *Zeitschrift für Psychologie,* **202,** 349–377.

Custance, D. M., Whiten, A., & Bard, K. A. (in press). Can young chimpanzees (*Pan troglodytes*) imitate arbitrary actions? Hayes and Hayes (1952) revisited. *Behaviour.*

Fodor, J. A. (1992). A theory of the child's theory of mind. *Cognition,* **44,** 283–296.

Gallup, G. G., Jr., & Suarez, S. D. (1986). Self-awareness and the emergence of mind in humans and other primates. In J. Suls & A. G. Greenwald (Eds.), *Psychological perspectives on the self* (Vol. **3**). Hillsdale, NJ: Erlbaum.

Gopnik, A. (1993). How we know our minds: The illusion of first-person knowledge of intentionality. *Behavioral and Brain Sciences,* **16,** 1–14.

Gopnik, A., & Meltzoff, A. N. (1986). Relations between semantic and cognitive development in the one-word stage: The specificity hypothesis. *Child Development,* **57,** 1040–1053.

Gopnik, A., & Meltzoff, A. N. (in press). *Words, thoughts and theories.* Cambridge, MA: MIT Press.

Johnson, D. B. (1982). Altruistic behavior and the development of the self in infants. *Merrill-Palmer Quarterly,* **28,** 379–388.

Leslie, A. M. (1994). ToMM, ToBy, and agency: Core architecture and domain specificity in cognition and culture. In L. Hirschfeld & S. Gelman (Eds.), *Mapping the mind: Domain specificity in cognition and culture.* New York: Cambridge University Press.

Lewis, M., & Brooks-Gunn, J. (1979). *Social cognition and the acquisition of self.* New York: Plenum.

Lewis, M., Sullivan, M. W., Stanger, C., & Weiss, M. (1989). Self-development and self-conscious emotions. *Child Development,* **60,** 146–156.

Povinelli, D. J. (1993). Reconstructing the evolution of mind. *American Psychologist,* **48,** 493–509.

Povinelli, D. J. (1994). How to create self-recognizing gorillas (but don't try it on macaques). In S. Parker, R. Mitchell, & M. Boccia (Eds.), *Self-awareness in animals and humans.* Cambridge: Cambridge University Press.

Povinelli, D. J. (1996). Chimpanzee theory of mind? The long road to strong inference. In P. Carruthers & P. Smith (Eds.), *Theories of theories of mind.* Cambridge: Cambridge University Press.

Povinelli, D. J., Bierschwale, D. T., Reaux, J. E., & Čech, C. G. (1996). *Do adolescent chimpanzees understand attention as a mental state?* Manuscript submitted for publication.

Povinelli, D. J., & Cant, J. G. H. (1995). Arboreal clambering and the evolution of self-conception. *Quarterly Review of Biology,* **70,** 393–421.

Povinelli, D. J., & Eddy, T. J. (1996). Chimpanzees: Joint visual attention. *Psychological Science,* **7,** 129–135.

Povinelli, D. J., & Eddy, T. J. (in press). Factors influencing young chimpanzees' recognition of "attention." *Journal of Comparative Psychology.*

Povinelli, D. J., & Godfrey, L. R. (1993). The chimpanzee's mind: How noble in reason? How absent of ethics? In M. Nitecki & D. Nitecki (Eds.), *Evolutionary ethics.* Albany: State University of New York Press.

Povinelli, D. J., & Povinelli, T. J. (1996). Review of *Mindblindness: An essay on autism and theory of mind,* by Simon Baron-Cohen. *Trends in Neuroscience,* **19,** 299–300.

Povinelli, D. J., & Preuss, T. M. (1995). Theory of mind: Evolutionary history of a cognitive specialization. *Trends in Neuroscience,* **18,** 418–424.

Povinelli, D. J., Zebouni, M. C., & Prince, C. G. (1996). Ontogeny, evolution, and folk psychology. *Behavioral and Brain Sciences,* **19,** 137–138.

Premack, D. (1988). Minds with and without language. In L. Weiskrantz (Ed.), *Thought without language.* Oxford: Clarendon.

Premack, D., & Dasser, V. (1991). Perceptual origins and conceptual evidence for theory of mind in apes and children. In A. Whiten (Ed.), *Natural theories of mind.* Oxford: Blackwell.

Premack, D., & Woodruff, G. (1978a). Author's response. *Behavioral and Brain Sciences,* **1,** 616–629.

Premack, D., & Woodruff, G. (1978b). Does the chimpanzee have a theory of mind? *Behavioral and Brain Sciences,* **1,** 515–526.

Savage-Rumbaugh, E. S., Murphy, J., Sevcik, R. A., Brakke, K. E., Williams, S. L., & Rumbaugh, D. M. (1993). Language comprehension in ape and child. *Monographs of the Society for Research in Child Development,* **58**(3–4, Serial No. 233).

Tomasello, M., Savage-Rumbaugh, E. S., & Kruger, A. C. (1993). Imitative learning of actions on objects by children, chimpanzees, and enculturated chimpanzees. *Child Development,* **64,** 1688–1705.

Whiten, A., Custance, D. M., Gómez, J. C., Teixidor, P., & Bard, K. A. (1996). Imitative learning of artificial fruit processing in children (*Homo sapiens*) and chimpanzees (*Pan troglodytes*). *Journal of Comparative Psychology,* **110,** 3–14.

CONTRIBUTORS

Daniel J. Povinelli (Ph.D. 1991, Yale University) currently directs the Division of Behavioral Biology at the University of Southwestern Louisiana (USL)–New Iberia Research Center and the USL Center for Child Studies. His research interests center on evolutionary analyses of psychological development in monkeys, apes, and humans. He is the recipient of a National Science Foundation Young Investigator Award and the 1994 American Psychological Association Distinguished Scientific Award for an Early Career Contribution to Psychology in the area of animal learning/biopsychology.

Timothy J. Eddy (Ph.D. 1992, State University of New York at Albany) is the current recipient of the Chimpanzee Cognition Post-Doctoral Fellowship at the USL–New Iberia Research Center. His current research interests concern the behavioral, psychological, and physiological correlates of interspecific animal relationships.

R. Peter Hobson (Ph.D. 1989, Cambridge University), a fellow of the Royal College of Psychiatrists, works at the Tavistock Clinic and University College London as Tavistock Professor of Developmental Psychopathology in the University of London. His interest in the development of interpersonal understanding involves a special research focus on early childhood autism. He is also a psychoanalyst. He is the author of *Autism and the Development of Mind* (1993).

Michael Tomasello (Ph.D. 1980, University of Georgia) is professor of psychology at Emory University and affiliate scientist at the Yerkes Regional Primate Research Center. His research interests focus broadly on processes of communication, social cognition, and cultural learning in children and nonhuman primates—with a special focus on the language acquisition of children and the gestural communication of chimpanzees. He is the author of *First Verbs: A Case Study of Early Grammatical Development* and the coauthor (with Josep Call) of the forthcoming *Primate Cognition.*

STATEMENT OF EDITORIAL POLICY

The *Monographs* series is intended as an outlet for major reports of developmental research that generate authoritative new findings and use these to foster a fresh and/or better-integrated perspective on some conceptually significant issue or controversy. Submissions from programmatic research projects are particularly welcome; these may consist of individually or group-authored reports of findings from some single large-scale investigation or of a sequence of experiments centering on some particular question. Multiauthored sets of independent studies that center on the same underlying question can also be appropriate; a critical requirement in such instances is that the various authors address common issues and that the contribution arising from the set as a whole be both unique and substantial. In essence, irrespective of how it may be framed, any work that contributes significant data and/or extends developmental thinking will be taken under editorial consideration.

Submissions should contain a minimum of 80 manuscript pages (including tables and references); the upper limit of 150–175 pages is much more flexible (please submit four copies; a copy of every submission and associated correspondence is deposited eventually in the archives of the SRCD). Neither membership in the Society for Research in Child Development nor affiliation with the academic discipline of psychology are relevant; the significance of the work in extending developmental theory and in contributing new empirical information is by far the most crucial consideration. Because the aim of the series is not only to advance knowledge on specialized topics but also to enhance cross-fertilization among disciplines or subfields, it is important that the links between the specific issues under study and larger questions relating to developmental processes emerge as clearly to the general reader as to specialists on the given topic.

Potential authors who may be unsure whether the manuscript they are planning would make an appropriate submission are invited to draft an outline of what they propose and send it to the Editor for assessment. This mechanism, as well as a more detailed description of all editorial policies, evaluation processes, and format requirements, is given in the "Guidelines for the Preparation of *Monographs* Submissions," which can be obtained by writing to the Editor, Rachel K. Clifton, Department of Psychology, University of Massachusetts, Amherst, MA 01003.